Cattle in the Cold Desert

Cattle in the Cold Desert

James A. Young and B. Abbott Sparks

▲▲ UNIVERSITY OF NEVADA PRESS / RENO & LAS VEGAS

Cattle in the Cold Desert was originally published by Utah State University Press in 1985.

University of Nevada Press, Reno, Nevada 89557 USA
New material copyright © 2002 by University of Nevada Press
All rights reserved
Manufactured in the United States of America
Design by Carrie House
Library of Congress Cataloging-in-Publication Data
Young, James A. (James Albert), 1937–
Cattle in the cold desert / James A. Young and B. Abbott Sparks.
p. cm.
Originally published: Logan, Utah : Utah State University Press, © 1985.
ISBN 0-87417-503-8 (pbk. : alk. paper)
1. Beef cattle—Great Basin—History—19th century. 2. Ranchers—Great Basin—History—19th century. 3. Ranchers—Great Basin—Biography. 4. Ranch life—Great Basin—History—19th century. 5. Range ecology—Great Basin—History—19th century. 6. Sparks, John, 1843–1908. 7. Harrell, Jasper, 1830–1901. 8. Grazing—Environmental aspects—Great Basin—History—19th century. 9. Great Basin—History—19th century. I. Sparks, B. Abbott, 1919–   . II. Title.
SF196.U5Y68 2002
636.2'00979—dc21   2002008742
The paper used in this book meets the requirements of American National Standard for Information Sciences—Permanence of Paper for Printed Library Materials, ANSI Z39.48-1984. Binding materials were selected for strength and durability.

11 10 09 08 07 06 05 04 03
5 4 3 2

*To the Memory of John Sparks, Cattleman, and
to those who followed on to the New Lands*

CONTENTS

ILLUSTRATIONS

PHOTOGRAPHS

FIGURES

ACKNOWLEDGMENTS

I greatly appreciate the many hours Glenda Eskstrom and Jamie Peer spent typing the many drafts of this manuscript. Appreciation is also expressed to Guy Rocha, Nevada State Archivist, and Phil Earl, Nevada Historical Society, who were always ready to help track down details and sources. The interlibrary loan staff of the Life and Health Sciences and Getchell Libraries, University of Nevada, spent many hours obtaining material used in this manuscript. I am equally appreciative of the many other librarians and individuals who provided answers to my numerous requests.

Several individuals with the Agricultural Research Service of the U.S. Department of Agriculture encouraged and prompted the development of this manuscript. They include Dr. Raymond A. Evans, Research Leader; Dr. Edward B. Knipling, former Area Director; and Dr. Peter H. Van Schaik, former Associate Area Director. Howard Sherman contributed both his encouragement and his skill as an editor to this project.

Newton and Andy Harrell of Twin Falls, Idaho, provided a valuable link between the present and such past historical characters as Henry Harris. The firsthand accounts of these two cowboys sparkle with the flavor and color of the glory days of the Sparks-Harrell ranching empire.

I am very grateful to Dr. Charles S. Peterson, Utah State University, for believing in the potential contribution of *Cattle in the Cold Desert* to western history. Special thanks to Alexa West and Linda Speth, Utah State University Press, for taking a chance on a different kind of history manuscript.

Most of all, I thank my wife, Cheryl, and my children, Theresa, Patrick, and Nancy Yini, for their patience and understanding.

JAMES A. YOUNG

I wish to express my sincere appreciation to Jim Young for his invitation to join him in this work, which he conceived and researched throughout his eminent background. He and his many associates in Nevada, and the intermountain ranching states, were a pleasure to work with.

My invaluable executive assistant over the last 47 years, Leta Lassetter, devoted over a year to the compilation of background data, copy and editing to assist in the development of this manuscript. Don Fickert and Laura Jacobus made significant contributions to graphics and copyediting.

Russell E. Bidlack, editor of the *Sparks Quarterly*; Mrs. Grey Golden, former Texas State Archivist; and Mary Sparks Matthews, editor and publisher of *Fourteen Frontier Families*, added valuable inputs as did Leland Sparks of San Francisco, California, and Nancy Sparks Lawrence of Twin Falls, Idaho.

The Matthews book, *Fourteen Frontier Families*, traces the Sparks and related families through three hundred years of their pioneering progress through eleven states following their arrival on this continent in 1660 as forefather generations to brothers John and Tom Sparks.

Other Sparks family background was contributed by B. Abbott Sparks Sr. and Van J. Sparks of Pauls Valley, Oklahoma (both great-nephews of John Sparks), as well as Estha Scoggins of Georgetown, Texas.

The many personal contributions to the completion of this work were inspiring and indispensable. I join Dr. Young in his tribute to Dr. Charles Peterson, and add my special appreciation to the endeavors of Sandy Crooms and her staff at the University of Nevada Press in bringing to life this third printing (expanded edition) of *Cattle in the Cold Desert*.

Newton and Lida Harrell, formerly of Twin Falls, Idaho, provided a living link with historical characters in the book. Newton, a graduate of the University of Oregon, who spent his life "cowboying," was a live-wire link to activities in the heydays of the Sparks/Harrell ranching empire.

Phil Earl and Guy Rocha were of invaluable assistance in obtaining and making available much of the material and photographs used in this manuscript.

B. ABBOTT SPARKS

On Cattle and Cold Deserts

All things occur within a larger context. Behind the individual events or circumstances is the grand mosaic of surrounding events—the geologic history, the human history, the environment.

Thus, around the people and land of the Great Basin there is a hint, a flavor, of events that have greater significance when taken as a whole rather than as individual occurrences or circumstances. Against the backdrop of the cold desert's sagebrush/grasslands is the pageant of man and his herds. It is easy from a twentieth-century perspective to try to discern the underlying "meaning of things," in both a scientific and a historical sense, but recognizing those specific events as meaningful when they were under way would have taxed anyone's intuition and intellect. Thus, it is doubtful that the early pioneers, ranchers, and sheep men realized the significance of events as they were actually happening. This difficulty is perfectly understandable; even today we have no perspective when events are actually happening—when their place in the larger context has not yet become clear.

However, as we try to point out in this book, if we are attentive we can begin to gain a very useful sense of the larger context even while the events are in progress. We can do this by sensitizing our awareness to fragments of history and science that we frequently ignore. These fragments are like puzzle pieces that, when recognized and assembled, give an understanding of the greater meaning of things, the larger context.

Thus, we have taken the disciplines of science and history with their individual events and circumstances, placed them in a specific place for a specific time, and woven them together to create a larger view of the Great Basin. The chapters alternate between land and man, but in a sense they are inseparable, and that is the main thrust of this volume—the larger picture of man and his cattle in the cold desert.

Another purpose of this volume is to provide perspective on the influence of grazing animals on the ecology of the sagebrush/grasslands. The scope of our coverage is designed to provide background information for those who earn their livelihood from grazing animals on sagebrush/grasslands, for professionals who manage such lands, and for individuals living in or outside the sagebrush/grasslands environment interested in the quality of life in this ecosystem.

This volume traces the history of man and his herds and flocks of domestic animals as they exploited the forage resources of a pristine environment. We do not tell the entire story—only the introduction and expansion period from roughly 1860 through the end of the nineteenth century. Essentially, this volume consists of a discussion of scientific principles and philosophies set in the context of historical events. It is not meant to be a biography of principal characters such as John Sparks and Jasper Harrell; however, it uses the life experiences of these and other pioneers to illustrate how man, the herdsman, interacted with the sagebrush/grasslands environment.

The story of the exploitation of the grazing resources of the sagebrush/grasslands recapitulates what has happened over virtually the entire surface of the earth except for the bleakest Arctic wastes and densest tropical jungles. The difference is that the sagebrush/grasslands was one of the last great vegetation resources to be suddenly, radically,

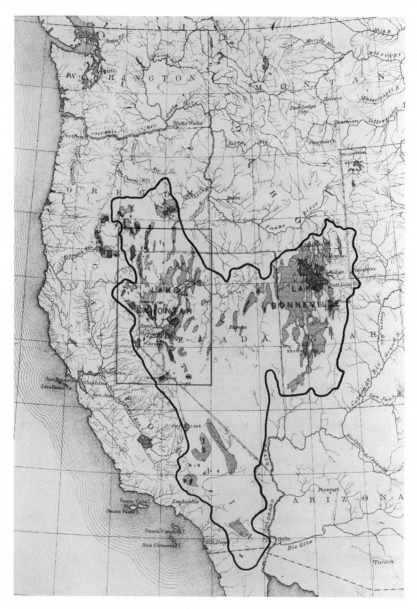

Figure 1. Boundary of the Great Basin. From I. C. Russell, *Geological History of Lake Lahontan* (U.S. Geological Survey, 1885).

and irrevocably changed by the introduction of domestic livestock. Because it is a relatively recent event, the process of the development of the livestock industry in the sagebrush/grasslands can be reconstructed with relative clarity.

Environmental quality is very much in the eyes of the beholder. A major objective of this account is to provide perspective for decisions on the nature of environmental quality in the sagebrush/grasslands. It is difficult for one to witness the degradation of one's own homeland, especially if the natural rate of environmental change is nearly static. Environmental changes in the sagebrush/grasslands tend to reflect sudden catastrophic events followed by long struggles toward equilibrium.

To provide the reader with a sense of urgency regarding such catastrophic environmental events, we introduce this volume with a prologue, an account of five violent days in August 1964. The events it describes occurred after the time period (1860–1900) on which we concentrate, but they were so dramatic that for a brief instant, a portion of the general public questioned environmental quality in the sagebrush/grasslands and related it to human actions such as the introduction of large numbers of domestic cattle almost a century earlier.

Halfway through this volume we relate another environmental catastrophe—the hard winter of 1889–90. Both disasters forced herdsmen, and the general public as well, to look at the consequences of their actions. For the general public, unfortunately, such perceptions are usually fleeting and easily lost in the haze of pseudostability that seems to follow sagebrush catastrophes.

The vegetation of the pristine sagebrush/grasslands was rather simple and thus extremely susceptible to disturbance. The potential of the environment to support plant and animal life was limited by lack of moisture and often by accumulations of salts in the soil. The native vegetation lacked the resilience, depth, and plasticity to cope with concentrations of large herbivores. When faced with grazing herds of cattle, the plant communities did not adapt; they shattered. This fact tends to make a review of grazing in the sagebrush/grasslands a horror story, a compilation of examples of what should not have been done. In perspective, the development of ranching in the sagebrush/grasslands was a grand experiment initiated by men willing to venture beyond the limits of accepted environmental potential to settle the Great Sandy Desert between the Rocky, Sierra Nevada, and Cascade Mountains.

The dynamics initiated by the introduction of domestic livestock into the sagebrush/grasslands are still under way. Each plant community that composes an effective environment or habitat type in the sagebrush/grasslands is a biological measure of the interacting factors that make up the environment. The plant communities also mirror a century of disturbance. Interpretation of the influence of history on potential is essential to developing management schemes for the environment. There are no more pristine sagebrush/grasslands to exploit. We must learn to constructively use and restore the existing resource or sink into an endless spiral of further degradation.

Five Days in August

At 10 A.M. on the Saturday morning of August 15, 1964, Elko, Nevada, was enjoying a pleasant summer morning with a few cumulus clouds drifting lazily east toward the towering bulk of the Ruby Mountains. The Ruby Mountains still supported patches of snow in sheltered, north-facing glacial cirques. Elko was hot and dry. Tourists felt sure they were going to die while driving across the baking Nevada desert. Highway 40, the major transcontinental route, passed through Elko and followed the Humboldt River valley southwest to the sink where the river terminated near the Carson Desert.

This was the pioneers' route in the great rush to California in the 1850s.[1] Modern travelers welcomed the comfortable motels and hotels of Elko after the heat of the desert. The electric excitement of the casino made them forget the endless sagebrush valleys and mountains. Liquor was cheap and the Elko nightlife was out of proportion to the size of the town, a transportation and supply center for the surrounding livestock ranches. Divided by two sets of transcontinental railway tracks, Elko County was advertised as the leading range livestock county in the nation. Some of the houses on the east side of the tracks had red lights over their doors and cross-country diesel tractors and semi-trailers parked in front with the motors idling. When the nightlife quieted down in the early morning, the exhausted tourists dropped into bed, only to be bounced right back out by the vibrations of a long freight train highballing through town.

Elko was a western town—a town with a cowboy heritage. The doctor, banker, barber, and undertaker dressed in western-cut suits or Levis and high-heeled cowboy boots. The restaurants and casinos were decorated in plush western motifs.

The lifestyle of Elko County ranchers had improved greatly since the end of World War II. Beef and wool enjoyed a seller's market after price controls were lifted. According to an article in *Life* magazine, the only thing in short supply in Elko in the late 1940s was one-hundred-dollar bills.[2] Many of the old ranchers sold to outside investors after the war. Bing Crosby bought a ranch on the North Fork, and Jimmy Stewart bought the old H-D Ranch on Thousand Springs Creek when the giant Utah Construction Company was sold. The H-D was the jewel in John Sparks's nineteenth-century ranching empire. Ranches became popular tax shelters for the new rich in the postwar era. Cattle prices remained relatively low during the early sixties, and very few ranchers were making a valid return on their capital investment.

As that August Saturday progressed, the Weather Bureau alerted the local fire-suppression agencies—the U.S. Forest Service (USFS) of the U.S. Department of Agriculture (USDA), the Bureau of Land Management (BLM) of the U.S. Department of the Interior (USDI), and the Nevada Division of Forestry (NDF)—of a 40 percent chance of thunderstorms.[3] The BLM had the most land to worry about; 67 percent of the land area of Nevada was under its control.

Bob Carroll, the BLM fire control officer in the Elko District Office, was worried. He had recently been successful in getting his fire crew increased from seventeen to twenty-seven men. He had started the summer with no standby fire crew. On June 27, three days after Carroll finally got his standby crew, they had their first fire. His fire crew was a collection of boys just out of high school, college students, and a few Indians from Duck Valley and the South Fork of the Humboldt. What they lacked in experience they made up in enthusiasm. They soon got experience, too. By late July they had fought seventy-eight fires, equal to the total number of wildfires for 1963, and double the number for some previous years.

Range managers had realized for thirty years that much of the lower-elevation sagebrush/grasslands were severely overgrazed. There were millions of sagebrush-dominated acres with virtually no herbaceous understory. The previous summer a BLM official from the Nevada headquarters in Reno had told a USDA range scientist that these vast acreages of degraded sagebrush range were an asset rather than a liability, an environment frozen in time. They were fireproof because they lacked herbaceous vegetation to carry fire. The BLM developed plans and obtained funds to selectively transform these degraded communities into grasslands by plowing them and seeding exotic (i.e., deliberately introduced) perennial crested wheatgrasses.[4]

The spring of 1964 had been prolonged and wet. The alien (i.e., accidentally introduced) weed cheatgrass responded to the late spring moisture by producing a tremendous crop of herbage. On the northwest side of Emigrant Pass in July 1964, range scientists clipped four thousand pounds of cheatgrass per acre from experimental plots. On normal years cheatgrass often produces only a few hundred pounds of herbage per acre, and on dry years virtually nothing. Cheatgrass is an annual that dries in early summer and provides a finely textured flash fuel that carries wildfire through sagebrush stands. The nine large fires that occurred during July showed the extent to which this accumulation of cheatgrass constituted a fire hazard. One of those fires burned forty-five thousand acres.

From August 1 to August 15 there were lightning storms somewhere in Nevada nearly every day. On July 29 there were flash floods at Yerington, Nevada, that tumbled cars around on the highway like rubber toys. On August 11 Tonopah, Nevada, had a similar storm. Elko expected at least one intense lightning storm each year. The fire-suppression officials could only hope that the lightning storm and peak fire conditions did not coincide.

Bob Carroll waited and worried. His boss, Clair Whitlock, the Elko BLM district manager, had left on vacation the night before, and Bob had the lonely feeling that he was sitting on a powder keg with a lit fuse. He checked the light plane he planned to fly for spotting lightning strikes and talked to the two Torpedo Bomber-Medium (TBM) pilots the BLM had on contract at the Elko airport. These pilots flew aging World War II navy bombers modified for dumping a slurry of bentonite (a form of clay) on fires. Pilots had to love flying and danger to be involved in this business. Some years there were no fires and they spent a lot of time waiting around hot airports, tinkering with worn-out airplanes. If the pilot was lucky, some fire agency paid him a standby fee.

The standby fire crew in Elko viewed the 40 percent prospect of lightning with mixed emotions. The pumpers were serviced and the bulldozers were loaded on the lowboy trailers and ready to go. The summer fire-crew members welcomed fires because they brought relief from boredom and petty jobs around the BLM yards. Most of all, fires meant overtime and extra dollars. On the other hand, it was Saturday night, and the pleasures of Elko were awaiting their youthful enthusiasm. Working hard and playing hard were equal parts of being on fire crews.

There had been so many fires during July that the professional range managers in the BLM had been forced to drop their regular jobs administering grazing on the national resource lands to concentrate on fire suppression. Fire was a physical thing that could be attacked and defeated through hard work and planning. Good land management was a nebulous thing with shifting goals, endless red tape, and insufficient funds that provoked anything from apathy to outright hostility from ranchers.

Neither Bob Carroll nor the pilots and standby crews knew it, but their fate was sealed by 10 A.M. that Saturday morning. An upper-level low-pressure system moved across central California in the early morning, and at 10 A.M. it was located south of Lake Tahoe over the Sierra Nevada.

At 1 P.M. there were towering cumulus and cumulonimbus clouds across the northern Great Basin. Winnemucca, Nevada, reported lightning and a sprinkling of rain in the early afternoon. Teleprinters clattered to life in Weather Bureau offices, and fire radios crackled with the ominous news that the cloud base was six thousand to nine thousand feet high, indicating a lack of moisture in the clouds. From Winnemucca to Wendover, Nevada, dry lightning cracked in sheets and massive single bolts on the tinderbox rangelands. The dry lightning flashed without accompanying rain.

By late Saturday afternoon, about thirty-five fires had been reported. One experienced BLM fireman started for the Willow Creek fire in a pickup, and on the way put out an estimated three hundred acres of spot fires that no one had reported. The TBMs flew into the thunderstorms at incredible risk and helped knock down sixteen of the fires. The remaining fires burned together to form the Boulder, Maggie Creek, Willow Creek, Palisade, and New Corral fire complex with the Upper Clover as a single fire. Someone coined the word "firestorm." It fit the situation and it stuck.

Bob Carroll luckily reached the district manager, Clair Whitlock, at his first vacation stop. He told Clair he had to return to Elko; they had a firestorm. As the district manager put the phone down to face his family with the news that the vacation was aborted, he wondered what in the hell a firestorm was.

Considering the enormous acreage burning, the suppression crew did quite well on Sunday with the limited manpower and equipment available. Ranchers organized their own crews to fight spot fires and generally cooperated with the government agencies in controlling the range fires. They were concerned about cattle in the fire areas and worried that the areas of dry forage they planned to graze that fall were being burned. The citizens of Elko were less concerned. It was the peak of the tourist season, hay was being cut on the ranches, and construction projects were in full swing. Anyone who wanted to work had a job, and no one was interested in helping the government fight fires. "Let it burn, it will improve the range," was the oft-repeated comment in street-corner conversations.

Rancher Joe Peretti, of the 7 Lazy Y Ranch near Elko, was frantic about his cattle. Normally cattle were quite capable of getting out of the way of wildfires, but these animals had never experienced a firestorm on sagebrush rangelands. Lurid tales appeared in the Reno and Salt Lake papers suggesting that hundreds or even thousands of cattle had been killed and that the loss of fall forage would force early sales and cost millions in losses.

It was a long, sleepless Saturday night for the BLM personnel. Their problem on Sunday was to find bodies to press into fire crews. The state BLM office in Reno responded quickly with help from other districts. The Humboldt District took over the Kelly Creek fire that was near their boundary. The state director of the BLM, J. Russell Penny, did not fit the stereotype of a successful bureaucrat. He was dynamic and could act decisively and organize effectively, but the only men immediately available from outside the bureau were winos off Lake Street, which at that time was the skid row of Reno. They volunteered quickly enough, but their physical condition made them a safety hazard when they reached the fire lines.

By Sunday night, August 16, outside fire crews and overhead personnel were arriving in Elko and being dispatched. The overhead personnel were supervisors present to take over the organization and direction of fire-suppression efforts. This was supposed to give Bob Carroll a chance for a few minutes' rest, but he could not rest easily knowing there were field crews who had worked twenty-four hours straight without rest or food.

The Nevada Youth Training Center, a school for incorrigible boys committed by the courts, had dispatched its fifty-man crew under NDF foresters. During the 1930s and the days of the Civilian Conservation Corps (CCC), it was discovered that tough kids could be molded into fire crews if they were given quality leadership, training, and esprit de corps. The practice may have had no lasting reform influence on the boys, but at the time, dispatchers referred to fifty *men* from the Youth Training Center.

The BLM fire boss at the Elko headquarters had accumulated a reserve crew by midday on Monday. The crew had been fed and loaded on buses, and was going out the gate of the BLM yards when word was received of a large new fire near Elko. This was the Sherman fire, and the buses were diverted to attack the blaze. About one-half hour later the Grindstone fire was reported. Clair Whitlock would later comment that the Sherman and Grindstone fires broke the back of the fire-suppression program. Not only did these two new sleeper fires exhaust all his reserves, but the conditions that caused them to explode spelled disaster. The temperature was rising, the humidity was dropping, and the afternoon winds were increasing.

The fire-fighting air force had been greatly augmented by Monday. The fire dispatcher allotted seven tankers to hit the Sherman fire when it was first reported. The fire was only seven miles from the airport, but despite the advantage of a short ferrying distance, the planes could not suppress it. The rate of its spread was beyond the experience of any of the firefighters; an estimated ten thousand acres burned in two hours.

As the sun went down Monday night, there was a glow west of Elko. The particularly aromatic smell of sagebrush smoke was evident even in the air-conditioned casinos. The attitude of the townspeople underwent a sudden change. The BLM was besieged with volunteers. An army of firefighters poured into Elko from throughout the West. Nevada State Forester George Zappettini called on Governor Grant Sawyer for help. The Nevada National Guard was called out to assist in fire suppression, crew feeding, and transportation. The Nevada Air National Guard took aerial photographs of the burned area. Senators Howard Cannon and Alan Bible called on Secretary of the Interior Stewart Udall, who promised to render all possible aid. Air Force bases Stead and Hill responded with medical teams, transportation, and more bulldozers.

Communications problems haunted the firefighters. The Boulder fire boss had fifteen aerial tankers dropping slurry in crossing patterns, with only one communications channel to direct all of them. There was not enough radio equipment, and much of what was available failed to function. Crews left Elko with drivers who had only a vague idea about where they were supposed to deliver the firefighters. A cook on the Boulder fire asked the fire boss to radio for six hundred rations for dinner; later he raised it to one thousand, then two thousand.

Wildfires in sagebrush vegetation normally explode during the late afternoon and quiet down after sunset when the winds die down. On Monday night, August 17, 1964, temperatures stayed relatively warm. The humidity was low, and the dawn came with swirling winds. The Boulder fire camp had to be moved several times during the night as the winds continually changed direction and rolled flames over the camps.

Tuesday was the day of the air force. More than forty aircraft were involved, including lead planes, tankers, and reconnaissance planes. There were twenty-one air tankers flying at one time. In addition, there was a large fleet of charter and U.S. Air Force planes. The runway was too short for the Air Force's largest planes, but c-119 cargo planes were unloaded as they rolled by without stopping. The Federal Aviation Administration (faa) set up a special visual flight rules control tower on top of the airport's terminal to handle the traffic. On Tuesday, August 18, Elko's airport handled more flights than Los Angeles International. Something relatively new in fire fighting was the use of fifteen helicopters to transport crews to remote locations with speed and ease. Fire crews still remember their relief when a helicopter brought fifteen gallons of badly needed drinking water. The fire air force became supertechnical when a usfs research plane carrying infrared sensing equipment capable of determining the extent of the fire through smoke was added to the fleet. Another research plane from the Desert Research Institute, University of Nevada, seeded thunderheads with silver iodide crystals.

The TBM pilots were making a dollar per minute—and earning every cent. They came in low through smoke and fire turbulence at ninety knots, below the cruising speed of their aging aircraft. A converted B-17 showed up at Elko with a thousand-gallon tank in its bomb bay. On its first run the entire load hit a concentrated area and dug a trench across the fire line. Depending on the size of the load, each slurry drop cost three hundred to five hundred dollars. On Tuesday, seventy-nine thousand gallons of retardant were dropped on the fires. One BLM fire observer's plane had a defective starter; he risked life and limb to turn the prop over by hand.

On Sunday, August 16, phones were ringing at 8 A.M. in BLM, National Park Service, and Bureau of Indian Affairs offices throughout the Southwest. The word went out—Elko, Nevada, had a huge fire and Indian fire crews from across the Southwest were needed. Fire bosses needed tough individuals who were accustomed to the outdoors and knew how to use axes and shovels to build fire lines. Southwestern land managers were familiar with the chronic unemployment on southwestern Indian reservations, and Indians had the qualities required of firefighters. But the local firefighters would have to accept them. Elko more or less fit the stereotype of western towns in its collective attitude toward Native Americans. Individually, the townspeople had many Indian friends, and there were many successful—by Anglo standards—Indian laborers, ranch hands, and professionals in the Elko area. On many of the larger ranches, local Indians were a significant part of the labor force and the haying crews. Although there were no signs in the plush Elko restaurants and casinos that said "Indians not welcome," however, the Indians who patronized the bars next to the railroad tracks knew they were not welcome uptown.

The Southwest Indian crews had already proven themselves as

firefighters on hundreds of fires. On the deadly Haystack burn in northern California they had held Indian Creek Ridge with hand tools when heavy equipment operators turned and ran.[5] But local papers often devoted more space to how the Indians danced for rain than to how skillfully they constructed and held fire lines.

Harold "Pete" Davis of Socorro, New Mexico, was a liaison officer with a twenty-five-man Taos Indian crew. The crew chief, Albert Martinez, rousted his men out of the Taos pueblo early Sunday morning on August 16. Edward Archuleta said good-bye to his wife, Cesarita, and five children, and left with the crew to fight fires in Elko County for $2.09 per hour. Air Force c-119s airlifted the Taos crew to Elko. Buses took them from the airport to a downtown tourist restaurant where they were fed a steak dinner. The waitress and townspeople in the restaurant were very glad to see them. After the meal, they were taken to the Elko High School football field, which was being used as a staging area, and then transported to the Palisade fire with a crew of Zia Indians.

The Navaho Number 1 crew from Rock Point Trading Post at Chinle, Arizona, had a more exciting experience getting to the fires. The crew was composed of eighteen-year-olds on their first fire and men in their late fifties who had toured the West fighting fires. When they arrived at the fire, no one knew where the fire lines were. They also lacked sufficient hand tools. After many delays, paper sleeping bags were distributed. There were not enough bags for the entire group, but they bedded down in the dust as best they could to await the dawn. Twice during the night, the camp was awakened by cries that the fire was coming closer and they had to move in a mad scramble. For sheer terror, a fast-moving wildfire approaching at thirty miles per hour in the middle of the night in strange country is hard to top.

The Indian crews were not the only ones on the move at the Boulder fire on that terror-filled Sunday night. The air force had sent tractors equipped with bulldozers to the fire with airmen, third class, operators fresh out of heavy equipment school. They had operated tractors through

obstacle courses on training grounds, but not in rugged terrain in the
dark with the danger of entrapment by fast-moving fires. The fire boss
put experienced Indian firefighters on the tractors beside the airmen to
offer advice and steady taut nerves. Some of the Southwest Indians had
limited English vocabularies, but they calmed the nerves of the young air
force men with a look, a nod, a hand on the shoulder.

Experienced wildfire fighters knew the characteristics of sagebrush
fires and tried to use this knowledge to their advantage. Fires burn up-
hill much better than downhill. Experienced firefighters never get
upslope of advancing fires, especially if there are rocks or cliffs ahead that
block their escape. Near Orvada, Nevada, there is a monument to the
memory of a crew of ccc men who failed to obey that rule. However, on
the Boulder fire the flames refused to obey the rules. Flames roared
upslope and crested the ridges. Then, instead of dying down as the standard script prescribes, the flames rolled down the back slope without
even pausing. Firefighters were astounded to see flames advance against
the wind through some trick of convection.

The convection column above the Boulder fire was so towering and so
intensely hot that it literally broke up thunderheads. The blazing walls
of flames distilled gases from the sagebrush fuel ahead of the fire itself.
Occasionally, these gases would explosively ignite, causing the flames to
leapfrog ahead in spectacular fashion. The Brewer's sparrow is a bird
species characteristic of degraded sagebrush rangelands. On the Boulder fire, firefighters reported seeing Brewer's sparrows being kicked up
from the sagebrush cover by advancing flames. The birds tried to fly away
from the oncoming fire, but the suction created by the updraft impeded
their progress until the zone of distilling gases caught up with them.
When the distillation zone ignited, the birds vanished in a puff of incan-
descent gases.[6]

Ross Ferris, the fire boss on the Maggie Creek fire, cussed the whirlwinds on Monday afternoon. I. C. Russell once described the whirlwinds
that start out on salt-flat playas and dance along as sensuous columns that

seem to support the entire summer Nevada sky.[7] The whirlwinds that
danced across the Maggie Creek fire were dirty with ashes and dust. They
picked up cow chips and pieces of sagebrush bark, some still smolder-
ing. When these embers hit unburned fuel, the fire was again off and
running.

The botanist P. B. Kennedy visited the area of the Maggie Creek fire
in 1901.[8] The plant communities of the Maggie Creek watershed he described were largely degraded at that time, but the alien cheatgrass had
not yet invaded the area. Now, sixty-three years later, cheatgrass provided the fuel that flashed the fire from shrub to shrub.

Marion Escobar had come from the Las Vegas office of the BLM to be
fire boss on the Palisade fire. He was pleased with the Zia and Taos Indian crews, who built fire trails so fast that fire bosses had a hard time
staying out of their way. Despite their maximum efforts, though, the new
fire raced over the hills at thirty-five miles per hour and eventually consumed thirty thousand acres. That night, Pete Davis's crew of Taos Indians were sitting in the dust eating dinner when a young hot shot crew
from the Boise Fire Control Center marched through shouting cadence
and stirring up the dust. The Indians turned to Davis to ask a collective,
"Why?"[9]

Back in town, residents of Elko were worried about the effect the fires
would have on the upcoming hunting seasons. Each fall thousands of
hunters, many of them from out of state, descended on Elko County to
harvest upland game birds and mule deer. It was big business for Elko.

On Tuesday the army of firefighters grew to more than three thousand
men, 260 vehicles, sixty tractors, fifteen helicopters, and forty fixed-
wing aircraft. Fifteen federal, state, and county agencies were cooperat-
ing. The fires appeared to be contained, but Nature was not yet finished.
Tuesday afternoon brought one of the strangest phenomena of the entire fire complex. Many people saw it, but no clear explanation for it has
ever emerged. At 4 P.M. on Tuesday afternoon the winds across the Boulder fire suddenly intensified, creating a rolling cloud of dust, ash, and

smoke flicked with flames that headed right for the Maggie Creek fire with the apparent intent of joining the two fires into one. To the Maggie Creek firefighters, it looked like the end of the world was rolling down on them at sixty miles per hour. Witnesses say that a "funnel cloud" hit the Boulder fire, after which confusion reigned. Crews were busy avoiding the flames while trying to see with eyes stinging from smoke, dust, and ashes. Smoke hid the fire from the planes circling overhead.

Fire headquarters in Elko received frantic phone calls from ranchers at Tuscarora, Nevada, that a new fire had broken out north of the Boulder fire. Crews were dispatched to reinforce a scratch force of ranchers. When the winds hit at 4 P.M., the Boulder fire made a spectacular run of five miles in ten minutes! It was mere chance that there was no one in its way. No one could have run or driven over the rough roads fast enough to escape.

The fires were finally contained by Wednesday, and the crews started picking up the trash and cleaning up the mess of three thousand men who had camped in the sagebrush. There were strong words expressed by northeastern Nevada ranchers and townspeople that something was wrong with the sagebrush environment. Why were firestorms rushing across the landscape? It had to be the government's fault. The old-timers said that it was never like this in their youth. There were many postmortem discussions among professionals about the logistics problems. As one Elko fire boss laconically put it, "If the darn fire had lasted five weeks instead of five days, we would have gotten command and communication problems solved."

All the forces involved mounted a dedicated response to the 1964 firestorms in Elko County. This amounted to emergency treatment of the symptoms without a postmortem search for the cause. Cattle had grazed in Elko County in large numbers for about ninety years before the 1964 firestorm.[10] During that period much of the pristine environment of the sagebrush/grasslands of western North America was irrevocably altered. What we know as ranching, or the cowboy culture, evolved during that time. Thick, juicy steaks, one of the products of ranching, came to symbolize luxury in American diets.

Was this agricultural development built on consumptive exploitation of the natural resources of the sagebrush/grasslands? Will the twenty-first century see the reduction of Nevada's sagebrush/grasslands to annual grass and weed ranges? Biologically, conversion to annuals ranks somewhere between feasible and probable. It can happen and probably will unless, as a culture, we collectively change our ways. Wildfires such as the 300,000-acre Elko fire complex in 1964 are the triggering mechanisms that push sagebrush/grasslands ecosystems into dynamic successional changes. The flames of the wildfires destroy the shrubs that have frozen the degraded plant communities into sagebrush dominance and allow the herbaceous vegetation to respond. Wildfires such as the 1964 firestorms are not the cause of environmental degradation; they are the product. And the degradation of sagebrush rangelands is not the product of the activities of one resource consumer or of the land-management agencies; it is a product of our culture.

Ranching as we now know it was not transplanted to the sagebrush/grasslands; it evolved in place. Environmental impact statements are now written to predict the impact of grazing on national resource lands. Without a historical perspective as to how the sagebrush/grasslands arrived at their current ecological, social, and economic conditions, however, it is impossible to accurately prepare or interpret environmental impact statements.

After the ashes of the 1964 Elko fires had cooled, land management professionals initiated a crash program to try to rehabilitate the burned areas. But the townspeople, ranchers, and many professional land managers eventually lost interest in finding out what conditions in the sagebrush environment had brought fires to the edge of Elko. The problems have not gone away. The accidental combination of forage fuel and dry lightning will return. Through the application of appropriate technology, the environment can be restored to an approximation of the pristine environment with stable communities of forbs, shrubs, and grasses. If the burned sagebrush ranges are not restored, the alien weeds will inherit the sagebrush/grasslands, paving the way for repeated burnings and a continuing downward spiral of degradation.[11]

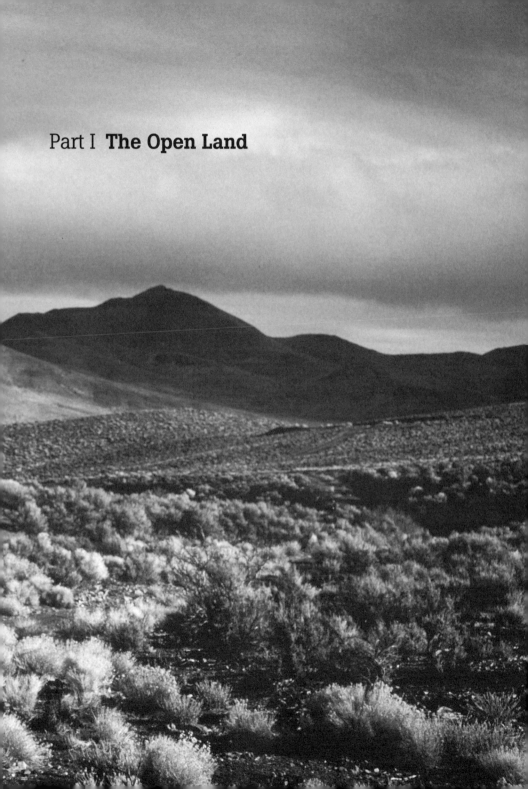

Part I **The Open Land**

There was not a single gate to open or fence to obstruct the movement of man and his herds between Salt Lake City at the foot of the Wasatch Range and Genoa at the eastern base of the Sierra Nevada. In the valleys of the central Great Basin, a person could ride for one hundred miles and never find a tree with sufficient shade to protect a rider and horse from the glare of the noonday sun. The barren salt flats reflected the dazzle of light and created mirages on which the bases of the mountains seemed to float. The mountains were islands in an arid desert sea, straining to reach up to the clouds that yielded water, the gift of life in this environment. The land awaited man.

CHAPTER 1

Gray Ocean of Sagebrush

The relentless silver gray reflection of the cold desert's sagebrush landscape could not help but impress those first few travelers who ventured
West. Oregon-bound travelers got their first taste of sagebrush near Fort
Laramie, Wyoming, and the gray ocean of sagebrush increased as they
proceeded westward. John C. Frémont had difficulty getting a wagon
through the dense stands of sagebrush on the Snake River Plains. The
perennial grasses that did exist within the sagebrush were mature, dry,
and harsh when the Oregon settlers reached the sagebrush plains in
early autumn.[1]

The endless uniformity of a sagebrush-dominated landscape tends to
blur differences and hide the details of this land's plant communities.
Sagebrush/grasslands are plant communities in which species of sagebrush form an overstory and various perennial grasses form the understory. In western North America, from southern Canada to northern
Mexico, a group of closely related sagebrush species comprise an endemic (i.e., occurring here only) section of the worldwide genus *Artemisia*.[2] They occur in varying amounts over 422,000 square miles in
eleven western states. Sagebrush/grasslands occur in all the western

states, but the discussion here focuses on the range plant communities of the Intermountain area.

The Intermountain area is the vast region from the Rocky Mountains on the east to the Sierra Nevada and Cascade ranges on the west. It is bounded on the south by the true warm deserts and on the north by coniferous forests. The Intermountain area is a cold desert—a semiarid to arid region where the winters are bitterly cold and often snowy.

How did the first settlers to enter this area view it? Anthropologists use the term "contact period" to indicate the period of first contact between indigenous cultures and European or American trappers, explorers, and settlers. Sources of information for the contact period in western North America are largely trappers' journals often written or edited after the actual contact in the field. Their major subject is the fur trade, with comments on the general environment usually secondary. Comments on plant communities are frequently no more than asides and often must be interpreted from other statements. Since trappers tended to travel and camp in river or stream valleys where beavers were likely to be found, most of the written comments concern these areas. The mountain ranges of the Great Basin tend to run north and south and are oriented in echelon, and travelers could avoid crossing them by going around them. Nineteenth-century travelers tended to group all shrubs in the Intermountain area as sage or wormwood even though the trails passed through greasewood- or saltbush-dominated landscapes. Essentially, one can use the records left by trappers and early explorers of the Intermountain region to confirm preconceived ideas of the pristine vegetation of the sagebrush/grasslands.

The extensive records kept by the early Mormon colonists provide the best account of the pristine sagebrush/grasslands environment. George Stewart, who was then with the Intermountain Forest and Range Experiment Station, summarized the available records in 1940. Stewart was uniquely trained and experienced for the role. After first becoming a successful agronomist and college teacher, he joined the Forest Service as a range ecologist. Stewart's position in the Mormon Church gave him access to the records of the early church colonies. It is interesting that Stewart takes quotations from the journals written during the contact period to emphasize the abundance of grass under pristine conditions, while T. R. Vale, who examined many of the same sources, concludes the opposite and stresses shrub dominance.[3] Generally, the records indicate that the abundance of shrubs increased as the Mormon settlers moved westward and southward in the Intermountain region until they passed into true desert vegetation.

Early travelers along the Oregon Trail had the opportunity to view a good cross section of the sagebrush ecosystem from Fort Laramie to the Columbia River. Their opinion of the sagebrush country was partially dependent on the time of the year they crossed the Snake River Plains.[4] The journals of those who crossed the area in late summer or early fall stress the sagebrush, lack of forage, and dust. Because of the time interval required for an overland journey from Missouri to Oregon, most travelers crossed the sagebrush/grasslands during the late summer, after they had traveled through the Great Plains, one of the world's foremost grasslands. Thus, the travelers crossed the plains during the peak of the growing season, and the sagebrush/grasslands obviously suffered by comparison.[5]

Anglo-American exploration and fur trading began in the Snake River area in 1809 and continued until 1846.[6] Most of the activities of the fur trappers were confined to the upper Snake River and adjacent areas. However, the hunt for beavers was much like the later prospecting for gold; the trappers followed virtually every stream in search of wealth. Peter Skene Ogden, who commanded the Snake River Brigade for the Hudson's Bay Company, left detailed and believable records of his extensive travels in the Intermountain region.[7] The Snake River Brigade traveled with two hundred horses used for riding and for carrying supplies, traps, and furs.[8] These animals depended on forage obtained along the line of march and in the vicinity of winter camps. The trappers who composed the brigade were avid hunters. When the brigade traveled in the eastern Snake River country where there were American bison, the hunters would exasperate Ogden with their wanton killing of game animals and failure to pay attention to the business of trapping. However, when the brigade was traveling in the Great Basin and south-central Idaho, forage for the horses was easy to find, potable water was scarce, and big game for food was difficult for Ogden's professional hunters to locate.

The initial settlements in much of western North America were established either at convenient points along transportation routes or at sites where important minerals were discovered. The Mormon settlements of the Intermountain area were an exception to this rule. The Mormons had to pick specific environments suitable for agriculture if they were to survive. Within thirteen years after Salt Lake City was founded in 1847, a series of outlying Mormon colonies had been established from Lemhi in Idaho to Genoa, Nevada, to San Bernardino, California. The sites for these settlements were carefully selected to include areas suitable for irrigation to support intensive agricultural and grazing lands capable of producing the meat, milk, and draft animals necessary for the colonists' survival. The extensive records kept by these early colonies provide the best descriptions of the environment available for the development of productive sagebrush/grasslands.

In the late eighteenth century, the American bison occasionally extended its range across the northern portion of the sagebrush/grasslands into northeastern California, the Malheur and Harney Basins of eastern Oregon, and even to the Columbia Basin. This is roughly the bluebunch wheatgrass portion of the sagebrush/grasslands. The American bison had withdrawn from northern Nevada long before historic times. Therefore, the Thurber's needlegrass portion of the sagebrush/grasslands had no concentrations of large herbivores under pristine conditions. In the early nineteenth century the number of American bison on the upper Snake River and Green River drainages increased as a result of hunting pressure east of the Rocky Mountains; after 1830, the populations west of the mountains were exterminated. The spread of trade that provided rifles to the Indians hunting for robes, promiscuous hunting by trappers, and several severe winters contributed to the bison's demise.[9]

The buffalo, or American bison, is the only large herbivore to exist in large numbers on the sagebrush/grasslands during recent geologic times.[10] At the close of the Pleistocene, the upper Snake River Plains were grazed by native species of mastodon, camels, horses, and ancestors of the American bison. All of these animals but the bison became extinct, and rabbits, rodents, and harvester ants became the major consumers in the sagebrush/grasslands. Certainly the pronghorn remained, but pronghorns were scarce in the Intermountain area and very scarce in the Great Basin. The Goshute Indians of Deep Creek in eastern Nevada practiced the communal activity of driving pronghorns with systems of traps, blinds, and barriers. Under pristine conditions the pronghorn populations in the various eastern Nevada valleys were sufficient to support only one of these drives each decade. It is no wonder the fur trappers had a hard time finding camp meat in this region.[11]

In contrast to the members of the deer family, which are fairly recent emigrants from Asia by way of the Bering Strait land bridge, the pronghorn is a true native of the sagebrush/grasslands of North America. The females bear their young in May, and the kids are soon following their mothers through the sagebrush. Bands of three to twenty animals are common in summer, and in winter the pronghorns collect in even larger bands. They are migratory in the sense that those in the higher elevations of the summer range move down to lower territory where there is less snow in winter, but essentially pronghorns always live in the sagebrush/grasslands.

Many of the rodent species that populate the sagebrush/grasslands are adapted to utilize metabolic water to satisfy their moisture requirements. They seldom, if ever, actually drink water, and instead obtain their moisture requirements from the food they eat. This gives them a tremendous competitive advantage over the large herbivores, which are limited to grazing within the range of infrequent waterholes. A second adaptation of many of the rodent species is the use of underground burrows, which moderate the environmental extremes of the Great Basin.

The black-tailed jackrabbit (*Lepus californicus*) is probably the most common consumer of plant material in many parts of the sagebrush/ grasslands. A traveler across Nevada sees more jackrabbits of this species than all the other small animals combined. The black-tailed jackrabbit is abundant in virtually all of the lower-elevation sagebrush areas. A black-tailed jack was spotted a mile out on the barren Fourteen-Mile Salt Flat east of Fallon, Nevada, and one was killed at 11,700 feet on Mount Jefferson in Oregon, illustrating the range of the species in the Intermountain area.[12]

Numbers of this species fluctuate so markedly that almost every Nevada resident has noted the phenomenon. The cause of the sudden crashes in rabbit populations may be tularemia, a bacterial disease caused by *Bacillus tularense*, which also attacks man. In jackrabbits, the mortality rate may reach 90 percent of populations. When the black-tailed jackrabbit populations are near their peak in a given area, they can be extremely destructive to crops, especially to irrigated fields in a generally sagebrush environment. Under pristine conditions, the white-tailed jackrabbit (*Lepus townsendii*) was probably much more abundant than the black-tailed jackrabbit.

Reconstruction of the pristine environment is not limited to historical records. Range ecologists are continually looking for relic areas where plant communities exist in equilibrium with the natural environment. Such areas remain because of natural "fencing" such as a mesa with sheer walls or lava flows that stock cannot cross; steep slope angles; or distance from stock water. In the sagebrush/grasslands water points are scarce and unevenly distributed across topography that is often rugged. This creates uneven utilization by grazing animals leaving areas long distances from water or on steep slopes ungrazed.

To look at the present sagebrush environment with an eye to differences in the past, one must understand the complex vegetation structure behind this gray landscape. It is necessary first to learn the identity of the major plant species, and then to recognize how the plant species fit together to form communities.

Within the Intermountain area there are two major subdivisions: the Snake River drainages and the Great Basin. The Snake River is part of the Columbia system, which also contains the Columbia Basin, a region that historically supported a northern extension of the sagebrush/grasslands. In southern Idaho, the Snake River flows into a deep canyon walled by nearly vertical basalt cliffs. On top of these cliffs are extensive undulating plains that were formerly clothed with sagebrush—the Snake River Plains.

Across a mountainous divide south of the Snake River Plains lies the other subdivision of the Intermountain area, the Great Basin. The Great Basin is a physiographic area with somewhat indefinite boundaries (see Figure 1). Roughly, it lies between the Sierra Nevada on the west and the Wasatch Mountains on the east, but its tributary valleys extend to Wyoming. To the southeast, it grades into high plateaus near the Colorado River. Southward, the province extends through the Mojave Desert of California in Baja California. Thus, the Great Basin includes most of Nevada and Utah with fringes in California, Oregon, Idaho, and Wyoming. The name "Great Basin" was first applied by Frémont in 1844 when he scientifically established that no water drained into the ocean from this huge area.

The plant communities of the sagebrush/grasslands are a measure of the potential of the environment (see Figure 2). The history of the exploitation of the sagebrush/grasslands is also reflected in the present plant communities. The walls of flames exploding through the big sagebrush/grasslands of Elko County in 1964 were a vivid expression of this history.

Pristine plant communities in equilibrium with their environment are adjudged to be in excellent *range condition*. As the plants that com-

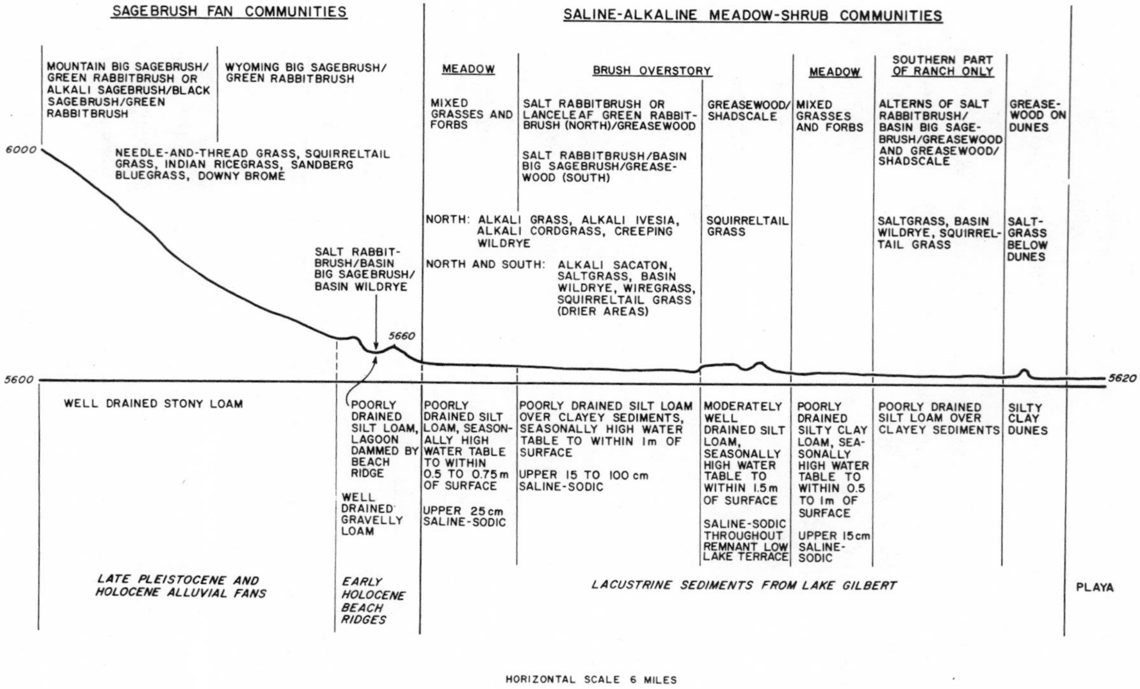

Figure 2. Sagebrush communities distributed on alluvial fans. From J. A. Young and R. A. Evans, eds., "The Physical, Biological, and Cultural Resources of the Fund Research and Demonstration Ranch, Nevada" (USDA/SEA. *Agricultural Reviews and Manuals*, ARM–W–11, June 1980). Used with permission.

pose the pristine community change in abundance—with, for example, the unpreferred shrubs increasing and the desirable perennial grasses decreasing—range condition drops to good, fair, or poor. The direction range condition is proceeding—either downward toward a degenerated condition or upward toward equilibrium—is called *range trend*. Range condition and trend are the basic concepts used to evaluate the impact of grazing animals on the environment. Obviously, knowledge of the pristine environment before domestic livestock was introduced is vital to establish condition and trend standards.[13]

There are many kinds of sagebrush growing in the sagebrush/grasslands, but the species that generally characterizes the environment is big sagebrush, *Artemisia tridentata.* This scientific name was first given in 1841 by Thomas Nuttall to plants collected from the plains of the Columbia River.[14] Big sagebrush is normally an erect shrub three to six feet tall, although dwarf forms occur occasionally. The trunk of the shrub is definitely woody with a stringy, fibrous bark. The silver gray hairs on the leaves and new twigs give the entire plant a light gray-green appearance. The light color of the leaves reflects much of the sun's incoming radiation and protects the plants from desiccation in the arid environment. The leaves persist through the winter, and when the current year's growth begins in early spring, it is difficult to distinguish it from previous seasons' growth. Flower heads appear in midsummer, with flowering in late August and September. The yellowish flowers are borne in clusters on flower stalks. The brownish black seeds begin to fall in October, and some persist until spring.[15]

Three characteristics of big sagebrush are especially significant to its grazing ecology: (1) this landscape-dominant shrub does not resprout when the aerial portion of the plant is burned in wildfires; (2) the species is composed of many ecologically distinct subspecies that in appearance or morphology are difficult to distinguish; and (3) the essential oil content of the herbage of big sagebrush inhibits the growth of the rumen microflora in cattle and, to various degrees, other ruminants.

The fact that big sagebrush does not sprout after being burned in wildfires has fundamental significance in the ecology of the species. The role fire played in the pristine environment of the sagebrush/grasslands is difficult to assess. In forests, the frequency of past fires can be determined by examining fire scars left on the trunks of trees. These scars, or "cat faces," indicate the frequency of past fires that damaged the cambium layers in the annual rings of the trees.[16] In the sagebrush/grasslands, however, there are no trees to record the frequency of fires. Under pristine conditions, wildfires eliminated the landscape's dominant shrubs and essentially released the native perennial grasses from competition. The native perennial grasses mature more slowly than the exotic invader cheatgrass. Thus, the fire season for pristine sagebrush/grasslands must have occurred in late August and early September rather than the mid- to late-summer fire season seen with cheatgrass.

After the big sagebrush has been consumed in a wildfire, the community that reoccupies the site is not devoid of shrubs. A number of shrubs that are subdominants to big sagebrush resprout from their roots or crowns after being burned. One of the important subdominant shrubs in big sagebrush communities is low rabbitbrush, a highly variable species with many distinct subspecies that sprouts from dormant crown buds. Almost all of the subspecies and forms of low rabbitbrush are spurned by large herbivores. Horsebrush, which sprouts from roots, is toxic to browsing animals. White-skinned animals become photosensitive after consuming its herbage. There are several shrub species that occur occasionally in big sagebrush communities and sprout after being burned. Species of plum such as desert peach, green ephedra, and ribes have extensive underground stems or woody crowns called lignotubers that protect buds and store food for regeneration after burning. Green ephedra is an interesting species because it is a gymnosperm, more closely related to the pine trees than to the sagebrush species. Its green, broomlike twigs, which it bears rather than leaves, make it one of the few vividly green species in this gray environment.[17]

Big sagebrush's susceptibility to wildfires leaves the character of the landscape open to sudden change by a single catastrophic event. It takes from ten to fifteen years for big sagebrush to reinvade areas where it was destroyed by fire. If the fire hopped and skipped through the stand, leaving many shrubs untouched, the rate of return is much faster. During the ten- to fifteen-year period after sagebrush burns, the sites are dominated by the root-sprouting shrubs mentioned above.

How frequent were wildfires in the pristine sagebrush/grasslands? Henry Wright determined that the interval between fires had to be greater than ten to fifteen years. If this were not true, rabbitbrush/horsebrush/grasslands would have been the pristine condition rather than sagebrush/grasslands.[18]

Big sagebrush characterizes the sagebrush/grasslands, but there are important subdivisions within the species. To recognize them it is necessary to compare the plants' morphology; chemical differences as identified by thin layer chromatography; cytological evidence from the number, shape, and characteristics of the chromosomes; ecological characteristics of the communities where the shrubs are found; and distribution patterns. Four subspecies are currently recognized: (1) basin big sagebrush, (2) Wyoming big sagebrush, (3) mountain big sagebrush, and (4) subalpine big sagebrush. Generally, mountain big sagebrush occurs at higher elevations, Wyoming big sagebrush occurs on drier alluvial fans, and basin big sagebrush is found on alluvial soils of the Great Basin, Snake River Plains, and innumerable mountain valleys.[19]

The third characteristic of big sagebrush that has great significance to cattle husbandry in the sagebrush/grasslands is the essential oil content of the herbage. The protein content of big sagebrush herbage approaches or exceeds that of cultivated alfalfa, and the branchlets have no ridges or spines to prevent browsing. On the surface, big sagebrush would thus appear to be an excellent forage species. No vertebrate is capable of breaking down and digesting highly lignified-cellulose plant material. The major herbivores, both wild and domesticated, capable of consuming and digesting coarse grasses are all ruminants. And that is where the problem with big sagebrush arises.

The rumen is a modified digestion system that provides anaerobic (oxygen-free) sites for the growth of microorganisms. The microorganisms break down the woody plant material into components that can be digested by the host animal. The volatile fatty acids that are the end products of rumen bacterial fermentation are the ruminant's major source of energy. Cattle can consume small amounts of big sagebrush with no problem. If big sagebrush herbage constitutes a relatively large portion of the diet, however, the activity of the rumen microflora is retarded or inhibited.[20] The same is true for mule deer and elk, although these native big game animals can consume more sagebrush than cattle. The only native large herbivore that makes big sagebrush a large portion of its diet is the pronghorn.

The inhibition of rumen microflora has been linked to the essential oils in the sagebrush herbage. These volatile oils are composed of terpene compounds and their derivatives.[21] The amount of volatile oils that individual big sagebrush plants contain depends on the season, the site where the plant is growing, the environmental stress to which the plant is subjected, and the subspecies involved. Mountain big sagebrush generally has a lower content of volatile oils than basin big sagebrush. There are probably inherent differences in the quantity of volatile oils among strains of big sagebrush as well, which raises the possibility that plant breeders could produce a big sagebrush that does not inhibit the microflora of cattle and mule deer.[22] The volatile oils found in big sagebrush may affect browsers even before the plant is eaten. Many large herbivores apparently base their forage selection on smell. Apparently, the strong-smelling essential oils in the herbage inhibit browsing.[23]

Once big sagebrush plants become established, they occupy a site for a very long period. There is a poor correlation between the size of sagebrush plants and their age. Big sagebrush plants with more than two hundred annual growth rings have been reported. The huge, treelike big

sagebrush plants that are occasionally found along drainage ways growing on old meadow soils are seldom more than seventy or eighty years
old. Their age probably reflects the time when the overgrazed meadows
were desiccated by deepening channels, and brush species invaded.[24]

Not all of the sagebrush found in sagebrush/grasslands is big sagebrush. In the western United States, there are about 422,200 square
miles of sagebrush. Of this area, approximately 226,370 square miles is
big sagebrush. The next most important species is silver sagebrush,
53,200 square miles; followed by black sagebrush, 43,300 square miles;
and low sagebrush, 39,100 square miles. In the Humboldt Basin of
northern Nevada, 45 percent of the landscape is dominated by species
of sagebrush: 40 percent is big sagebrush and 5 percent is low sagebrush.
Sagebrush species may be successional dominants in pinyon/juniper,
mountain brush, and mountain conifer communities in addition to the
45 percent of the landscape that is true sagebrush/grasslands. In northeastern California, the sagebrush/grasslands are roughly 50 percent low
sagebrush and 50 percent big sagebrush.[25]

Silver sagebrush is often associated in the Intermountain area with
the fine-textured soils and seasonal flooding of old lake basins. Silver
sage is restricted to the east side of the Sierra Nevada and eastern Oregon
and does not extend down into the highly saline/alkaline soils in the
depths of the Intermountain deserts. Silver sagebrush does sprout after the aerial portion of the shrub is removed; however, because of the
nature of the habitat it occupies, it is seldom burned in wildfires. The
various subspecies of silver sagebrush are widely represented in the
northern Rocky Mountains and the major river valleys on the eastern
slope of the Rocky Mountains.

Black sagebrush and low sagebrush are the major components of the
group of sagebrush species aptly named "low sagebrush." These species
are usually one-third the height of big sagebrush. Although similar to big
sagebrush in appearance, they often occupy very different landforms.
Low sagebrush is usually on the oldest landform in a given landscape

where soils have a well-developed clay horizon close to the surface. Clay soils take up water very slowly, and low sagebrush flats are extremely wet and sticky in the spring. Conversely, when dry, the soils are baked brick-hard.[26]

Low sagebrush is highly preferred over big sagebrush by big game and domestic animals. Sheep prefer black sagebrush for their winter browse. On the margins of the Carson Desert in western Nevada there are extensive stands of black sagebrush. After a century of winter use by sheep, these monospecific communities look like they were tended by a host of Louis XV gardeners. Each shrub is perfectly molded in a ground-hugging exotic shape.

Often only twelve to eighteen inches high, low sagebrush spreads over the landscape in a tidy gray film quite unlike the uneven texture of big sagebrush landscapes. Because they are old landforms, low sagebrush flats often have biscuit-and-swale topography. The biscuits have shallow, mounded soil profiles, and the swales may be devoid of soil with only stringers of frost-sorted rocks present. Plant communities called "balds" occur on high, exposed ridges. These communities are dominated by low sagebrush shrubs scarcely four inches tall that are shaped by winter gales.

Throughout the range of the sagebrush/grasslands there are a series of plant communities delineated by the dominant shrub species and the understory grass species. To those unfamiliar with them, this network of species may appear to be a bewildering array of variability. In fact, however, the plant communities are repetitive and easily identifiable. Recognizing them is important because they are phytometers, or living measurements, of a given local ecosystem composed of soil, topography, climate, animals, and the plants themselves.

The best-known big sagebrush community is big sagebrush/blue-bunch wheatgrass, which dominates in the Columbia Basin, eastern Oregon, much of Idaho, and extends down the higher elevations into the Great Basin. This is a bunchgrass community in which the dominant grass occurs in distinct bunches as opposed to a continuous carpet. In a bunchgrass community, much of the soil surface is exposed; only 20–30 percent of the area is covered by the canopies of plants.

To people more familiar with the humid East or the grasslands of the Great Plains, the sagebrush/grasslands with their bunchgrasses appear poverty-stricken. However, the bunchgrasses are in equilibrium with their environment. Not only is precipitation limited in quantity throughout the sagebrush/grasslands, but the timing of the precipitation is largely out of phase with temperatures suitable for plant growth. Most of the precipitation falls as snow during the winter months. Plant growth occurs during the spring and is limited by cold temperatures in early spring and by the exhaustion of soil moisture in early summer. Into the brief spring growth period are crammed all the functions of growth and reproduction. Even in the semiarid grasslands of the Great Plains, moisture falls during the summer, when temperatures are optimum for growth. Ten inches of spring and summer precipitation on the Great Plains is much more effective than ten inches of winter precipitation in the Great Basin.

Bluebunch wheatgrass is just as much a landscape-characterizing species as big sagebrush. On steep slopes, good stands of bluebunch wheatgrass resemble ranks of soldiers marching up the slope. Found hand-in-glove with bluebunch wheatgrass, but growing on more mesic situations such as north-facing slopes, is Idaho fescue. At lower elevations, especially in the southern part of the Intermountain area, Thurber's needlegrass replaces bluebunch wheatgrass as the main understory species in big sagebrush communities. In virtually every big sagebrush community, the short-lived perennial grasses squirreltail and Sandberg bluegrass are minor components. When the communities are disturbed, these short-lived grasses rapidly respond to renew the community.[27]

Several other species of perennial grass form communities with sagebrush species as well. Communities of needle-and-thread grass often dominate areas with sandy soil. All of the needlegrass species have long awns on their seeds. When they are in fruit, the various species of needlegrass can be relatively easy to identify by the shape and hairiness of the awns. Needle-and-thread grass carries this to an extreme. The seed itself is needle-sharp at its tip, and the four-to-six-inch twisted awn is a coarse, barbed thread.

Black and low sagebrush duplicate the big sagebrush communities in forming communities with the major perennial grasses. There is a low sagebrush/bluebunch wheatgrass community, low sagebrush/Idaho fescue, etc., and the same series exists for black sagebrush.

The perennial grasses begin growth in early spring. Most species flower in early summer and set seeds that are mature by midsummer. As they start growth in the spring, the perennial grasses mobilize food, stored as carbohydrates, for energy to support their growth. If the grasses are heavily grazed year after year and not given a chance to rebuild their carbohydrate reserves, they will die. If the native perennial grasses are heavily grazed every spring, they never have a chance to produce seed for new plants.

One thing that was missing from the pristine sagebrush/grasslands was native annual grasses. Six-weeks fescue (*Vulpia octoflora*) is a diminutive native annual grass that briefly graces the sagebrush range in the spring, but it is not a good competitor. The herbaceous communities had no vigorous, highly competitive annual grass capable of rapidly renewing communities destroyed by grazing animals. Many of the valleys of central and northwestern Utah that are in the big sagebrush zone had the aspect of true grasslands with dominance by bluebunch wheatgrass. Without question, various woody species of sagebrush, and especially big sagebrush, existed in all these environments. What is different in the present environment is the proportion of grasses versus shrubs.

Sagebrush communities are not usually colorful. The season after high-elevation sagebrush communities burn in wildfires, however, the renewing community can be a riot of color. Large forbs like arrowleaf balsam root and mules ears, members of the sunflower family, produce big, showy flower heads with bright yellow rayed flowers. Mixing with the yellow of the daisylike coarse forbs are the delicate red shades of the Indian paintbrush flowers. The roots of Indian paintbrush often parasitize the roots of sagebrush plants, robbing them of nutrients. The deep blue to purple colors of low larkspur spikes add radiant but deadly beauty to the gray shrub landscape. The larkspurs are poisonous to livestock, as is the death camus, a member of the lily family. Under the shrubs occur the delicate flowers of Chinese house.

One blooming plant picks a very specific habitat. The skunk monkey flower grows on the garbage dumps of harvester ant mounds. The harvester ants are important consumers in the sagebrush/grasslands. Virtually any oblique aerial photograph of a sagebrush community growing on an alluvial fan reveals a series of vegetation-free circles. These circles are the mounds of the harvester ants. The ants strip all the vegetation from a six-foot-diameter circle and build a small soil mound some eight to ten inches in the center of it. The busy ants harvest seeds from various plants growing in the sagebrush communities. Some of the seeds have glands that the ants relish. The ants take the seeds into their underground chambers, remove the glands, and then discard the seeds on the surface, creating a habitat for the skunk monkey flower.

The ability of the sagebrush/grasslands to support a cattle industry was something that had to be judged by mid-nineteenth-century agriculturalists before they risked their capital, and often their lives, by introducing livestock to those areas. Essentially, the would-be sagebrush rancher had no precedent by which to judge the potential of the sagebrush/grasslands environments, which were much more severe and undependable than the environments they had previously encountered in their spread across the country.

The pristine landscape that was available for exploitation by grazing consisted of a series of sagebrush/perennial bunchgrass communities.

This was not a paradise for grazers. The species of sagebrush that formed the overstories in these communities were not preferred by cattle as browse and if eaten in quantity would disrupt the animals' delicate symbiotic relationship with their rumen microflora. In contrast to the Great Plains, where great herds of American bison, pronghorn, and often deer and elk were displaced by the introduction of cattle, the sagebrush ranges of the Intermountain area had few large herbivores to displace. Most important, the herbaceous vegetation that composed the forage resource of the sagebrush/grasslands was not preconditioned for heavy utilization by large herbivore herds.

By the mid-nineteenth century, the grand experiment was about to begin—the exploitation of the last virgin grazing resource in the United States—the grazing of the sagebrush/grasslands.

CHAPTER 2

The Exploitation Pageant

The earliest rancher in the Intermountain area is a nebulous figure. He may have been a fur trapper or a freighter who overwintered in this area and grazed his animals. Whoever that nebulous figure was, he participated in the first exploitation of the region's grazing resources.

The time period 1860–1900 encompasses both the classical range operation period, in which cattle were allowed to roam freely on unfenced native ranges with minimal husbandry, and the period of ranch enterprise, in which livestock production was stressed and some fences and conserved forage were employed in the culturing of livestock. We define cattle ranches as agricultural enterprises that used extensive acreages of native vegetation for the production of livestock, with red meat the product marketed.

Nebulous figures provide little information for history books, so to illustrate the evolution of ranching in the sagebrush/grasslands we will introduce real cattlemen into our story. Woven together with the story of the land of the Great Basin are the stories of Jasper "Barley" Harrell, his son, Andrew J. Harrell, and their longtime business associate John Sparks. The activities of other ranchers and participants in this exploitation pageant are also incorporated into the narrative to illustrate specific points.

The Harrells and John Sparks are good examples of Intermountain ranchers for several reasons. The ranching enterprise they founded was one of the largest ever developed in the Intermountain area. It persisted as an operational unit until the mid-twentieth century, long after the deaths of its founders. Their ranching empire was centrally located in the Intermountain portion of the sagebrush/grasslands. Both families were experienced in the Spanish system of open-range livestock production practiced in Texas and California before they came to the sagebrush/grasslands. The Sparks family had a long association with new lands agriculture in the southeastern United States before they reached Texas. Such an association implies participation in livestock production on the woodland ranges of the Southeast. The ranching enterprise founded by these men was responsible for technological innovations, especially in water development and livestock breeding and husbandry, and therefore provides a good example of the nineteenth-century evolution of such processes. Finally, John Sparks was a dominant political figure of the time and, as such, left historical records of his activities.

The ranching empire that the Harrells and Sparks developed occupied portions of the three major drainage basins of the Intermountain area: the Lahontan, the Bonneville, and the Snake River. The history of the development of ranching in the Intermountain area tends to be segregated into these geographical areas. The Snake River sweeps in a great arc through southern Idaho, forming a huge valley bounded on the north by the mountains of the Idaho batholith and on the south by the mountainous lip of the Great Basin. In southeastern Idaho, the Bear River, a tributary to the Great Salt Lake, encroaches into the Gem State. To the west, the Owyhee River and its tributaries extend deep into north-central Nevada, draining the vast Owyhee Desert into the Snake. Between the Owyhee and the Bear are a series of streams that rise in the highlands at the border of the Great Basin and flow north to the Snake. On its rush to the Columbia, the Snake River flows through deep canyons with nearly sheer walls. On top of these walls and extending to the far horizon is a vast, empty plain of sagebrush, sand, and raw lava flows.

To the nineteenth-century traveler who did not know how to approach its vastness, Idaho was disguised in various ways, and in places appeared openly hostile. The early travelers' descriptions of the land were seldom favorable.[1] In his volume *Astoria,* for example, Washington Irving described the Snake River valley and its formidable appearance:

> A dreary desert of sand and gravel extends from the Snake River almost to the Columbia. Here and there is a thin and scanty her-bage, insufficient for the pasturage of horse or buffalo. Indeed, these treeless wastes between the Rocky Mountains and the Pacific are even more desolate and barren than the naked, upper prairies on the Atlantic side; they present vast desert tracts that must ever defy cultivation, and interpose dreary and thirsty wilds between the habitations of man, in traversing which the wanderer will often be in danger of perishing.[2]

The Snake River Plains is a generally flat area bordered by rugged mountains and extending in a concave curve to the north across southern Idaho. When viewed in the strong sunlight of early morning or in the lengthened purple shadows of evening, the plains appear to be absolutely level, and from many points of view, of ocean extent. In fact, the plain of the Snake River is a dissected plain. In some places its surface is mildly uneven and in others it is exceedingly rough owing to the character of the naked lava of which it is composed. It is also marked by many volcanic cones and by broad uplands formed by vent flows. The channels cut deeply into the Snake River lava, and the elevations that rise above the plain are minor features in comparison to the plain itself.[3]

The brilliant descriptions of the nineteenth-century geologist I. C. Russell are invaluable in reconstructing the pristine environment of the Snake River Plains. "One must become familiar with the characteristics," he observed,

. . . and learn to judge them by their own standards before their beauties are revealed. To the traveler from humid lands, where every hillside is clothed with verdure and every brook flows from a shadowy vale, they will at first seem repellent deserts, on which a long sojourn would be intolerable. When the sun is high in the cloudless heavens, the plains are gray, russet brown, and faded yellow, but with the rising sun and again near sunset they become not only brilliant and superb in color, but pass through innumerable variations in tone and tint. When the approaching dawn is first perceived, the sun is seemingly a great fire beneath the distant edge of the plain. A curtain was quickly drawn aside, revealing a limitless picture suggestive of the view a mariner sometimes has on approaching a bold coast while the actual shoreline is still below the horizon. The distant mountains, rising range above range and culminating in some far off sun-kissed peak, are the most delicate blue, while all below is dark and shadowy. As the sun mounts higher colors deepen, becoming violet and purple, of a strength and purity never seen where rain is frequent. Purple in all its rich and varied shades is the prevailing color imparted to arid lands when the sun is low in the heavens. As the dawn passes the light becomes stronger and the rich hues fade. The mountains recede and perhaps vanish in the all pervading haze, details become obscure even in the immediate foreground and the eye is pained by the penetrating light. The shadows, if canyon walls are near, are sharply outlined and appear black in contrast with the intense light reflected from the sunbathed surface. The light grayish-green leaves of the ubiquitous sagebrush give color, or perhaps more properly lack of color, to the plains and enhance their monotony. In the glare of the unclouded noontide summer sun the plains are featureless, or perhaps their expression is distorted and rendered grotesque or vague and meaningless by the deceptive mirage. During the cloudless summer the glories of sunset are on the earth, not in the sky. As the sun disappears, a well-defined twilight arch arises in the east, the shadow of the earth on the dust particles in the air. As evening approaches there is a gradual change from glare to shadow. The broad plain becomes a sea of purple on which float the still shimmering mountains. The shadows creep higher and higher, until each serrate crest becomes a line of light, margining rugged slopes on which every line etched through centuries by rills and creeks reveal its history. The mountains seemingly grow in stature and unfold ridges and buttresses separating profound depths. One marvels at the diversity and strength of the sculpturing of what but a few moments before appeared flat, meaningless surfaces. The flatland has details everywhere on its surface. The mountains stand boldly forth as sculptured forms of amethyst and sapphire, every line on their deeply engraved slopes, although leagues distant, clearly visible. The last ray of sun winks below the horizon and darkness descends quite suddenly. The cool, starlit summer nights are wonderfully magnificent, the heavens, without a cloud, are filled from horizon to zenith with stars which burn with a steady planetary light, such as is seen in our eastern humid lands only during clear winter weather.[4]

The Sawtooth Mountains of south-central Idaho express in their name their salient feature. Following the Idaho-Utah border, the mountains extend westward to the Jarbidge upland on the Idaho-Nevada border. West of the Jarbidge upland, the Owyhee Desert sweeps down into Nevada with a series of rugged dissected canyons in the desert plain. The Sawtooth Mountains, because of their geographic position and the environment they support, played an important part in the development of the area. The perennial streams that drain from the mountains supported habitat for beavers, which attracted fur trappers. Milton Sublette and a party of trappers from the Rocky Mountain Fur Company trapped along Goose Creek in 1832. The mountains provided an upland causeway for emigrants crossing the Great Basin that allowed them to avoid the Great Salt Desert to the south and a portion of the Snake River Plains to the north. The mountain environment provided summer ranges for livestock, and the streams provided natural meadows that could be enlarged by irrigation.

These pinnacled mountaintops are composed mainly of quartzite, but numerous granite spires and prominent limestone ridges are also sharply defined. The mountains rise boldly to heights ranging from six thousand feet to above ten thousand feet. At a distance the mountains appear to be a solid wall, but on entering their canyons it is possible to wander among and through the higher peaks by following a series of relatively low passes. It was by this series of passes that the Humboldt Emigrant Trail wound from Fort Hall in southeastern Idaho to the headwaters of the Humboldt and on to California.

Goose Creek is characteristic of the Sawtooth drainages. In this basin an outcropping of sedimentary rock on Trapper Creek contains numerous leaf fossils that date, on the basis of the deposit's stratigraphic sequence, to the Miocene epoch of geologic time some ten to fifteen million years ago. The Miocene was the beginning of the golden age of large herbivores when great herds of camels, horses, and mammoths grazed in western North America. Great rivers drained the present Great Basin into the ocean, and the Sierra Nevada and Cascade Mountains had not yet risen to rob the Pacific winds of their life-giving moisture and cast a rain shadow across the Intermountain area. The Trapper Creek fossil flora is composed of sixty-six species representing a forest in which both conifers and broadleaf hardwoods grew. A similar ecological situation today would be the upper Pacific coastal forest at three thousand feet elevation with its forty to fifty inches of annual rainfall. Some of the plants that compose the present sagebrush/grasslands evolved from those represented in the Miocene forest.[5]

The mixed coniferous forests represented by the Trapper Creek flora persisted under gradually drying conditions until the close of the Tertiary period. During the Pleistocene, the first epoch of our current Quaternary period, the Sierra Nevada and Cascade Mountains were uplifted, bringing extreme aridity to the Intermountain area with highly seasonal winter precipitation. This great change created the environmental conditions in which the sagebrush/grasslands ecosystem evolved. The Pleistocene epoch, or Ice Age, is very recent in geologic history, and therefore the vegetation formation of the sagebrush/grasslands evolved only recently and may never have reached equilibrium with the newly arid environment before domestic livestock were introduced.

Despite its forbidding appearance, the Snake River Plains did not go unclaimed. In 1837, endeavoring to keep American interests away, the Hudson's Bay Company purchased Fort Hall from New England merchant Nathaniel Wyeth. The fort became an outpost on the eastern perimeter of the company's Pacific division. Evidence of great pasture potential was scattered in every direction around Fort Hall, and settlers had only to notice the opportunity.[6]

In 1843, John C. Frémont camped southeast of Fort Hall on the southeast Bear River and wrote in his journal:

I can say of it, in general terms, that the bottoms of this river (Bear), and some of the creeks which I saw, now form a natural resting and recruiting station for travelers now and in all the time to come. The bottoms are extensive, water excellent, timber sufficient, the soil good and well adapted to grains and grasses suited to such an elevated region. A military post and civilized settlement would be a great value here, grass and salt so much abound. The lake will furnish exhaustless supplies of salt. All the mountains here are covered with a valuable nutritious grass called bunchgrass, from the form in which it grows it has second growth in the fall. The beasts of the Indians were fat upon it; our own found it a good substance, and its quantity will sustain any amount of cattle, and make this truly a bucolic region.[7]

At Fort Hall the Oregon Trail pioneers could choose an alternate route to happiness. If Oregon no longer had its original fascination, they could trade that dream for a new one in California. After 1848, the attraction of California and gold became so great that Hudspeth's Cutoff eliminated the need to go to Fort Hall by breaking away from the Oregon Trail near Soda Springs, Idaho. Diaries written by the pioneers include brief descriptions of the area. One traveler "passed over a valley covered with wild wheat as high as my shoulder. It was headed out and looked like a cultivated wheat field."[8] These reports indicated high country for summer range with water available and rich bottomlands with tall grass for winter grazing. In 1855 colonization moved into what was to become the Idaho Territory when a party of Mormons left Utah to establish a mission in the Lemhi Valley. Indian hostility forced the colonists to return to Utah. The Hudson's Bay Company abandoned Fort Hall in the mid-1850s, and the Snake River Plains remained in the hands of the Indians with occasional travelers hurrying through along the Oregon Trail.

In August 1860 E. D. Pierce found rich gold deposits on the Clearwater River in the Idaho Territory. By 1861, hundreds of miners were prospecting in the mountains. Several strikes were made within seventy miles of Fort Boise in 1862. Suddenly there was a reason for people to come to Idaho and a market for beef.[9]

In the spring of 1864 William Bryon was engaged in the business of butchering beef and feeding miners in the Boise Basin. He had imported five hundred beefs from the Washington Territory. With the onset of winter, he drove the animals south toward Nevada, hoping to find a mild valley with tall grass in which he could winter them. He planned to cross the Snake River on the ice, but when he reached the river, he found that it was not frozen. Stuck on the riverbank, Bryon and his herders suffered from the cold and wind, as did the cattle. Worried because he was trapped against the river in a desert environment with no feed available, he awoke the next morning to find the cattle "full as ticks" but was at a loss to explain what they had eaten. Watching the animals forage, he saw that they eagerly consumed the herbage of winterfat (*Ceratoides lanata*), a native shrub.[10] Undoubtedly, the value of winterfat has been rediscovered many times, but its ultimate importance in the development of the cattle industry cannot be overestimated.

Joseph Pattee was an agent for the Hudson's Bay Company at Fort Hall. He accumulated a considerable number of cattle by trading rested animals for sore-footed animals of emigrants on the Oregon Trail. After the company withdrew from Idaho, Pattee continued to accumulate cattle. In 1867 he decided to push westward to the Raft River valley and establish a new ranching area. His cattle wintered well in the valley.[11] Pattee's operation did not persist, but it showed that cattle could be wintered in the Raft River valley.

In 1865, near the point where Rock Creek leaves its canyon and ventures out among the rolling hills of the Snake River Plains, James Bascomb established a ranch and stage station. This venture extended the ranching frontier halfway across the Snake River Plains from Fort Hall. As with the first step into the Raft River valley, the Rock Creek Station was not destined to last. Bascomb was killed by Chief Buffalo Horn during the Bannock War. The Bannocks were probably the most warlike of the southern Idaho Indian tribes. The tribe was herded onto a reservation at Fort Hall in 1869, and this greatly encouraged the spread of ranching. In 1878 a small band of Bannocks stole forty head of beef from George Shoup, who was pasturing the cattle in Lemhi County in southeastern Idaho and killing the animals for jerked beef. The Bannocks moved off the reservation and went on a wide sweep across southern Idaho and eastern Oregon. Their actions became more hostile as they proceeded west wantonly destroying cattle and killing ranchers.[12]

The killing and destruction of stock hardened the settlers' attitude toward the Indians. Living with conflict, fear, and resentment, a great many early settlers developed harsh feelings about Indians. Typical of this rationale, which was passed from generation to generation, was the stance of pioneer Idaho cattleman David Shirk:

I want to say right here, and in all sincerity and with undue prejudice, that the only "good Indians" I ever knew were dead ones. As a
race, they are absolutely devoid of all feelings of gratitude, or any
kindred spirit. There may be, and probably are, exceptions but as a
race they are cruel, heartless, and treacherous as a coyote. Hu-
manitarians may shed their tears over the "poor" persecuted Red
Man, but mine are reserved for their victims, hundreds of whom
are heartlessly butchered by these painted and befeathered dev-
ils.[13]

If the early ranchers of Idaho had been restricted to sore-footed cattle
brought across the plains by settlers and occasional drives from the Oregon or Washington Territory, they would have needed years to fully stock
the vast ranges. Instead, the sagebrush/grasslands ranges were fully
stocked within a decade by thousands of cattle imported from Texas and
California. There was a demand for beef in southwestern Idaho in the
Silver City mining district. Pioneer cattleman Con Shea brought in Texas
cattle in 1868. He was on his way to Texas to buy cattle when he met two
drovers named Miller and Walters in the vicinity of the Raft River and
bought a trail herd they had brought from Texas. Miller and Walters returned to Texas and the next season trailed a herd of stock to the Bruneau
River valley. In 1869 J. G. Shirley and C. S. Gamble reached the Fort Hall
bottoms with three thousand head of Texas cattle. In the mid-1860s
Shirley had been ranching on the Fort Hall bottomlands when the U.S.
government bought the land for Indian reservations. In exchange for his
land Shirley received six sections of land on the Raft River at the mouth
of Cassia Creek. Later in the fall, attracted by the abundant forage, Shirley
and Gamble moved the herd to the Raft River valley. The next year Shirley
and his associates trailed ten thousand more cattle into the Raft River
area. The following year, 1871, A. D. Norton and M. G. Robinson proved
more venturesome and moved farther west to the Rock Creek area near
the present town of Hansen, Idaho.[14]

The bottomlands along the Raft River, Goose Creek, and Rock Creek were noted for their dense stands of tall native grasses. On the benches above the creeks grew extensive stands of winterfat. There were occasional tracts called "winter parks" that had abundant winterfat and natural hot springs with patches of willows for shelter.[15]

The early settlers noted that sudden winter snowstorms were usually quickly succeeded by thaws during the hours of sunshine. Frequently, warm "chinook winds" blew during winter. These winds caused snow to disappear as if by magic, leaving the plains and lower slopes of the bordering mountains bare.[16]

The first ranches were extremely crude. The rancher generally lived in a sidehill dugout and, at most, built a pole corral to hold his working horses. Lumber was not easily available and was expensive to import, and houses and other structures were built with whatever materials could be found locally.

George Miller spent the winter of 1867–68 ranging his cattle on the Snake River and hired David Shirk to help with the herding. The two men found a place near the river that had an abundance of feed and decided to locate their winter quarters there. They dug down into the riverbank about four feet, deep enough for a room which when framed in had proportions of ten by fourteen feet. A large willow ridgepole supported the roof and was covered with smaller brush and ryegrass. Six inches of dirt completed the roof insulation and provided weather protection. A chimney was constructed from rock and dirt, and a blanket covered the doorway. The house seemed adequate, for seldom did anything inside freeze.

Chinook winds often gave unexpected surprises to dugout dwellers. One winter evening when Shirk and Miller went to bed there was more than a foot of snow on the ground. During the night the wind began to blow. Toward morning Shirk was awakened by an unusual sound. He jumped from the bed, intending to look outside, and found himself standing in four inches of water. The sound that had awakened him was water running down the hill and into the dugout. For the remainder of the winter the two men did not have to wet the dirt floor to keep the dust down. Outside, the ground that eight hours before had been covered with snow was now bare.[17]

At the same time that the first ranchers were struggling to become established in Idaho, a parallel development was taking place in Nevada. The honor of being the first man to winter cattle in Nevada apparently goes to Captain H. A. Parker, a wagon master employed by Ben Holliday. In 1851 Parker wintered in the Carson Valley with a train of freight animals and three milk cows.[18]

The discovery and development of the Comstock Lode in 1861 created an enormous demand for beef. The first source of supply was California. The California ranchers were quick to see the advantage of ranching in western Nevada once a local market developed there. In 1864 a disastrous drought in California brought thousands of cattle to western Nevada. Rich mineral strikes at Austin in 1864, Pahranagat in 1865, White Pine in 1866, and Eureka and Pioche in 1866 brought new markets and spread stock growing across Nevada.

Up until 1858, all the cattle in Nevada belonged to the so-called American breeds. These were the common cattle brought by emigrants from the eastern United States. In 1858 two droves of Spanish Longhorns, the offspring of Spanish cattle that had been brought from Mexico to the Spanish missions in Alta California, arrived from California. Drovers Dorsey and Nottinger delivered fifteen hundred head to the Truckee Meadows, and L. B. Drexol brought two thousand head to the Carson Valley.

Occupation of the northern two-thirds of Nevada by ranching interests between 1860 and 1880 must rank as the most significant event in the agricultural development of the state. Ranchers, following the eastward movement of the mining frontier after the Comstock discovery, usually found their markets in the camps established after new strikes.

The building of the Central Pacific Railroad up the Humboldt River valley and across northwestern Utah to join with the Union Pacific forever changed the market for the livestock industry in Nevada. No longer were Nevada ranchers dependent on local mining camps, which required only a few animals every week. The completion of the railroad opened seasonal markets for large numbers of animals to feed the growing population of California or to ship to the midwestern livestock markets.

The Central Pacific Railroad continued from Wells, Nevada, to Tecoma, north of Pilot Peak on the Nevada-Utah border, and then across the Great Salt Lake Desert. The railroad route placed Thousand Springs Creek Valley within easy reach of a shipping point and also provided a transportation center for cattle driven from south-central Idaho.

The Humboldt River valley, from Big Meadows at its terminal sink to Elko, became the site of the base properties from which ranchers conducted extensive range operations. The Little Humboldt River draining the east slope of the Santa Rosa Mountains and flowing down through Paradise Valley to join the main river at Winnemucca supported another center of livestock production. The Reese River joins the Humboldt at Battle Mountain, and the Reese River valley extends into central Nevada for 150 miles from the Humboldt. On the upper reaches of the Reese, the high mountains of central Nevada provided another environment conducive to livestock production. At Carlin, Nevada, ranchers spread south into Pine Valley and north along Maggie and Susie Creeks. Near Elko the Humboldt River divides into three major forks. The South Fork, as its name implies, swings south to drain the southwestern face of the Ruby Mountains; the North Fork swings north to parallel the Independence Mountains; the Main Fork, or Marys River, continues northeast to head against the Jarbidge Mountains opposite the headwaters of the East Fork of the Bruneau River in Idaho. All of the forks of the Humboldt head toward mountain ranges with peaks at least ten thousand feet tall. Each of the valleys associated with the upper forks of the Humboldt proved to be exceptionally well adapted for cattle husbandry. In combination, they came to form Elko County, one of the most important range livestock production centers in the western United States.

The Humboldt River route of the California Trail has been called the "high road to the West," and rightly so. The trail left Soda Springs, twisted through the mountain causeway past the City of Rocks, and went down Birch Creek and into Goose Creek. The trail led up Goose Creek to the far northeastern corner of Nevada and then down Rock Creek to Thousand Springs Creek, which drains into the Bonneville Lake Basin. After following Thousand Springs Creek to its headwaters, the emigrants crossed the divide at the head of the creek to enter the Lahontan Basin and followed Bishop Creek down to Humboldt Wells.

Upper Goose Creek and Thousand Springs Creek are among the most isolated areas of the modern Great Basin, but in the 1850s they were on the mainstream of the western migration. Many people had the opportunity to see the land's potential to support domestic livestock, but most saw only the endless sagebrush and felt only fear of harassment from hostile Indians.

Thousand Springs Creek is similar to hundreds of other drainages in the Great Basin. It starts and ends in obscurity. The headwaters of the drainage are on Antelope Peak and Burnt Crown Mountain north of Wells, Nevada. At first the stream heads north as if striving for the headwaters of Salmon Falls Creek and eventual freedom in the Pacific Ocean. However, this is not to be, and the creek turns south to Bill Downing's H-D Ranch, where Toano Draw enters from the south. Once Thousand Springs Creek passes the H-D Ranch, it swings far to the north, avoiding Tony Mountain, and receives the waters of Rock Springs Creek and Crittendon Creek. The stream swings back south to finally emerge from the Toano Range into the valley at Montello. The total length of the creek is about fifty miles, with an average drop in elevation of eighteen feet per mile. The name Thousand Springs, an obvious choice considering the many springs that occur along the creek's course, was first applied on exploration and railroad survey maps prepared by Lieutenant Beckwith.

The valley at Montello has several significant features that contribute to livestock production. The oblong valley, some fifty square miles in area, is dominated on the southeast by the austere Pilot Range and the massive ten-thousand-plus-foot Pilot Peak. At the northeastern end of the valley there is a gap between the north end of the Pilot Range and the south end of the Tecoma Mountains. Through this gap extended an arm of pluvial Lake Bonneville, the Ice Age lake of which the Great Salt Lake is a remnant. On the vast alluvial fans that spread from the Pilot Range across the valley at Montello and through the northeastern gap to the Bonneville Salt Flats were thousands of acres of winterfat and other salt desert shrubs suitable for wintering cattle. A second feature of the area that enhanced livestock production was the several thousand acres of saline/alkaline plant communities that occurred where Thousand Springs Creek spilled onto the valley floor. These communities ranged from wet meadows and alkali bullrush (*Scirpus robustus*) marshes to extensive areas of the tall grass Great Basin wildrye (*Elymus cinereus*). This semiwetlands area provided an extremely valuable grazing resource that was much more productive than the sagebrush/grasslands. The last feature of the valley that enhanced livestock production was man-made: the Central Pacific Railroad.

Rock Creek and Crittendon Creek extend north from Thousand Springs Creek in broad basins with hundreds of thousands of acres of sagebrush rangeland. The Toano Basin extends south from Thousand Springs into the northern end of the Pequop Range and additional thousands of acres of rangeland, including extensive areas of pinyon/juniper woodlands. The town of Toano was an important freighting station before the Central Pacific Railroad was completed. Freight wagons left Toano for the McGill ranches to the south and the eastern White Pine mines. To the north, it was six days' fast freight to Boise City.

The first "ranch" between Salt Lake City and Carson Valley, Nevada, was located near Humboldt Wells and operated under dubious circumstances. Peter Haws, an early Canadian convert of Joseph Smith who fell out of favor with the Mormon settlers in the Salt Lake Valley, established himself on the Humboldt in 1854. Haws's daughter married Carlos

Murray, who settled in Thousand Springs Valley. The combined families raised a garden, sold the produce to emigrants, and traded fresh for sore-footed cattle. Dark rumors began to circulate that the Haws families were in league with Indians who stole back the animals they had traded to the emigrants. Eventually, Haws was forced to flee for his life, and Carlos Murray and his wife were killed in Thousand Springs Valley by the Indians they were accused of aiding.[19]

Probably the next rancher to try his luck in Thousand Springs Creek Valley was Bill Downing. Bill became disillusioned on the emigrant trail to California and dropped out to rest at a likely looking spring in the upper part of Thousand Springs Valley. The surroundings grew on him, and he decided to stay. He combined trading for sore-footed cattle with tending a garden and selling supplies he bought in Wells. He branded his stock with the H-D brand and called his operation the Ox-Yoke Ranch.[20]

In the late 1860s Bill Downing started to have neighbors, or at least someone living within a three-day horseback ride. Colonel J. B. Moore had commanded Camp Ruby in Ruby Valley on the east side of the Ruby Mountains during the Civil War. He used the troops under his command to help develop a ranch site. After the war, he was quick to purchase Texas Longhorns to stock his ready-made ranch. Colonel E. P. Hardesty had fought on the other side during the Civil War, but he also saw the opportunity for ranching in Elko County and brought cattle from Texas to land near Bishop Creek north of Wells.[21]

Large and small ranchers alike had found a fortunate combination of near-free grazing; easily monopolized water sources; growing markets in California and the Midwest; and a stable, easily controlled political climate sensitive to the requirements of an increasingly ranching-oriented economy.[22] In 1869 the editor of the *Elko Independent* wrote, "A new brand of business has been gradually growing up in raising and exportation of beef cattle—fattening of cattle and driving them to California has become an important business—large herds can be seen all along the valley of the Humboldt and its tributaries—In time, stock raising will be in sound value and extent greater than that of the production of precious metals."[23]

The time was right for the large-scale exploitation of the environment for livestock production. All that was needed was an entrepreneur with the skill to manage livestock, the capital to take advantage of the situation, and the guts to take the risk. Such a man was Jasper Harrell. Born near Augusta, Georgia, on August 16, 1830, Jasper grew up on the family cotton plantation under relatively wealthy conditions. In 1850 Jasper joined the rush to California, traveling by ship to Panama, across the isthmus, and then, again by ship, on to San Francisco. He mined at various locations in California and finally settled in Tulare County in 1856. Jasper developed an extensive ranching interest headquartered on Cross Creek near Visalia. Besides ranging cattle on the California annual range, Jasper Harrell brought thousands of acres under cultivation to cereal grain. He reinvested his money in real estate in the San Joaquin Valley in California. Harrell married Martha Bacon in 1857, and in 1861, a son, Andrew Jasper, was born to the couple. A. J. Harrell was later to play an important part in his father's ranching operations.[24]

Jasper Harrell was nicknamed "Barley" Harrell, supposedly because of the sack of barley he always carried behind his saddle to provide grain for his horse. His ranches in the San Joaquin Valley were producing thousands of bushels of barley, so he had a ready supply. Barley Harrell was very much a frontier figure. He rode the ranges of northeastern Nevada and south-central Idaho with a lever-action rifle in a saddle scabbard and a six-shooter in his holster. He paid his range crews every three or four months with gold and silver coins that he carried in his saddlebags. He timed his arrival at the cow camps for dinnertime. After dinner, Jasper would gamble with the crew and usually win back a large portion of the wages he had just paid them.

Jasper Harrell probably bought his first ranch in Nevada in 1870. He bought cattle in Texas in 1870 and had them delivered to Thousand Springs Valley. Among the cowboys employed to drive the herd was Louis

Harrell, Jasper's nephew from Georgia, who had decided to leave the post–Civil War South and take his chances with his uncle in the far West.[25] There was a ranch located at or near Tecoma belonging to a Mr. Armstrong which Jasper Harrell absorbed in a partnership.

Jasper Harrell was now well established on Thousand Springs Creek with extensive winter and spring/fall range available. But he still had no summer range, which required high-elevation mountains. The landforms close to Thousand Springs Creek lacked the necessary elevation. The topography and outline of the landforms of the Thousand Springs Creek Basin are the product of Tertiary volcanic action influencing extremely old Paleozoic and Precambrian sedimentary rocks. The mountains rise to an elevation of eight thousand feet, some three thousand feet above the valley floor. The ranchers established on the upper Humboldt were exploiting the Ruby Mountains, and those on the North Fork were using the Independence Mountains. Jasper Harrell sent his foreman, James E. Bower, north to upper Goose Creek in 1872 with a herd of three thousand cows to search for additional summer range. One day on a short exploring trip in the vicinity of Goat Springs, Bower climbed a high ridge and was able to see far to the north, and thus obtained his first glimpse of the Snake River valley. Bower could not resist riding down from the mountains to assess his new discovery. He rode down through pristine communities of bluebunch wheatgrass and Idaho fescue. The aspect was that of a grassland, not a shrub steppe. On reaching Cottonwood Creek, he met two Rock Creek cattlemen, A. D. Norton and M. G. Robinson. These two cattlemen were probably feeling lonesome and very vulnerable to hostile Indians at their remote ranch. They were quick to point out the potential of the country for stock growing. Most important, Norton told Bower of the wintering potential of the lower portions of the Snake River Plains. If Norton and Robinson had realized the growth potential of the cattle operation Bower represented, they might have been more subdued in singing the virtues of the south-central Idaho ranges. J. E. Bower hurried back to Nevada and told Jasper Harrell what he had seen. Harrell immediately started buying Texas cattle and pushing them north to establish a claim on the vacant range.[26]

Jasper Harrell's operation soon became established on Goose, Rock, and Salmon Falls Creeks, on both sides of the Nevada-Idaho boundary. One of his favorite ranches became the so-called Winecup field on Goose Creek, a clear, fast-flowing perennial stream meandering through emerald green meadows that contrast with the snow white volcanic ash bluffs that abruptly edge the valley. The foothills of the Goose Creek basin are partially clothed with pinyon/juniper woodlands. The round mass of Mahogany Butte rises to the northwest. The higher mountains support patches of lodgepole pine (*Pinus contorta*) and subalpine fir (*Abies lasiocarpa*).

The Winecup field is named for the ⴲ brand. There is no question that Jasper Harrell used this brand, but its origin is in doubt, being variously given as Wyoming or Texas. There is also the possibility that Harrell brought it from California. Harrell also used the Shoesole brand ⊂⊃, which was the fourteenth brand recorded in Elko County, registered on May 1, 1873. The Shoesole supposedly originated in Wyoming, where it was known as the Indian Moccasin. Jasper Harrell's horses were branded with a smaller version of the Winecup brand on the left shoulder. The earmark of the outfit was right ear cropped on top and left ear cropped under.[27]

The northern Nevada ranges were about to become more crowded. The winter of 1871 was extremely dry in California, and ranchers there drove thousands of cattle to Nevada. In addition, California was changing from an open-range area to dryland cropping of cereal grains during the early 1870s. In 1873 a fence law was enacted in that state making owners of livestock liable for damage caused by their animals. This spelled the end of the large open-range operations in the central valleys of California and sent many large ranchers to the Great Basin.[28]

The capital requirement for the type of expansion that Jasper Harrell engaged in during the 1870s was minimal. J. O. Oliphant estimated that the capital requirement in the Pacific Northwest for engaging in large-scale livestock production, outside of the cattle, was $1.12 per brood cow based on units of five thousand head. Jasper Harrell pumped the profits of almost two decades of successful ranching, grain farming, real estate investments, and banking in California into his Great Basin livestock enterprises.[29]

The range cattleman of the 1870s ran a frontier operation with very little capital. The open range was a practical proposition, for he could not buy the land necessary for the great herds to graze even if he wanted to. Most ranchers preferred to move on when the range became depleted. They were prepared to keep ahead of settlement, unhindered by ownership of depleted ranges.[30]

The failure of the great banking house of Jay Cook and Company precipitated a national depression in 1873. Jasper Harrell had the advantage of a rapidly expanding market to pull his range cattle operations through this depression. The silver mines of the Great Basin were booming, and the population of California was increasing rapidly. Because of its geographical position, history of settlement, and physical environment, the Great Basin was the last natural grazing land to be exploited in western North America—and, most probably, the world. The exploitation pageant now had its chief characters, and the saga continued across the sagebrush/grasslands.

CHAPTER 3

Left-Hand Trail to Hell

The cowboys from Texas who delivered cattle to Nevada must have thought they were riding down the left-hand trail to hell when they dropped the herds of Longhorns down into the arid basins left when Lakes Lahontan and Bonneville dried up. It would be difficult to find a more forbidding landscape in western North America.

Describing his travels in the Lahontan Basin in the 1880s, I. C. Russell noted:

> The valleys or plains separating the mountain ranges, far from being shady vales, with life-giving streams, are often absolute deserts, totally destitute of water, and treeless for many days' journey. . . . Many of them have playas in their lowest depressions— simple mud plains left by the evaporation of former lakes—that are sometimes of vast extent. In the desert bordering the Great Salt Lake on the west and in the Black Rock Desert of northern Nevada are tracts hundreds of square miles in area showing scarcely a trace of vegetation. In winter, portions of these areas are occupied by shallow lakes, but during the summer months, they become so

baked and hardened as scarcely to receive an impression from a horse's hoof, and so sun cracked as to resemble tessellated pavements of cream-colored marble. Other portions of the valleys become encrusted to the depth of several inches with alkaline salts which rise to the surface as an effervescence and give the appearance of drifting snow.[1]

In a humid environment, forage production depends on periodic rainfall during the growing season. As ranching operations spread across North America in the nineteenth century, ranchers encountered increasingly arid conditions. In the Intermountain area, precipitation was infrequent during the growing season. Would-be ranchers had to adapt to the physical and biological constraints of the sagebrush/grasslands environment. Permanent water, meadows, irrigable land, and the proximity of winter and summer ranges affected where ranches were located and which land was purchased.

Despite their abundance, the sagebrush/grasslands were far from a perfect place to produce red meat from cattle. The green feed period was short—spring and early summer. Just as calves born in the spring became independent grazers in midsummer, the native grasses became mature and their forage quality rapidly declined. To avoid this loss in forage quality, livestock followed the green feed up the mountains and grazed on high-mountain summer ranges.

Ranchers found winters in the Intermountain region much too severe for year-round grazing. As a result, transhumance (man and livestock moving with the seasons) agriculture developed in the area; sagebrush ranges were used in the spring and fall, mountain ranges in the summer, and salt desert ranges in the winter. Cattle were wintered on the margins of the salt deserts, and winter grazing was looked upon as very profitable.

After the Ice Age, Lake Bonneville was 19,750 square miles in area and had a maximum depth of 1,000 feet. The Great Salt Lake is a remnant of this lake. Lake Lahontan covered 8,422 square miles, and the deepest part, the present site of Pyramid Lake, was 866 feet deep.[2] The accumulation of soluble salts left in these old lake basins after the water disappeared earned them the name salt deserts. Not all of the soils in these basins are salty, but all of the basins receive very little rain—often four inches or less per year falls in the centers of the basins. Precipitation occurs during the cold winter months when temperatures prohibit plant growth.

Despite their apparent simplicity, playas play an important role in the ecology of Intermountain environments. Many of the playas are deflated (i.e., have their sediments blown away) during the dry portion of the year. Even during the winter months it does not take long for the surface of the playas to dry. The winter snow cover melts rapidly because of the surface's salt content. The clay-textured soil particles on the playa's surface flow together when wet, making it impossible for moisture to percolate into the soil. If a considerable amount of moisture accumulates or flows onto the playa surface, a shallow lake forms. Smaller amounts of moisture evaporate and leave crystals of soluble salts on the surface soil. The wet playa glistens, then turns dirty gray as the film of water evaporates. When the playa is completely dry, the salt crystals are so dazzling under a bright sun that it is impossible to look across the flats without one's eyes watering.

The crystal dazzle does not last very long, for soon the winds start to stir across the playa. The white playas are like huge mirrors reflecting the sun's energy back into the atmosphere. The reflected rays merge to mirror the sky and create shimmering mirages that offer false hope of water where there is none and cut off the bases of the towering mountains in the distance. Far out on the playa, the horizon is composed of spectral mountains that appear to float in space and continually retreat before an advancing traveler. The reradiated energy heats the atmosphere, and soon whirlwinds start to spiral across the playa surface. These swaying and bending columns, often two or three thousand feet high, rise from the plains like pillars of smoke and are characteristic features of the desert. The energy in these columns is surprising. More than one cow-boy has chased his hat across the salt flats.

At the end of a sunny, dry day, north winds blow across playas that have changed greatly in appearance. The white dazzle of crystals is gone, re-placed by a dull cream color. The salts have been redistributed in selec-tive patterns. Geologists who first studied the silver ore deposits of the Great Basin were impressed by the occurrence of silver chloride (or horn silver) on the surface, which they attributed to salt (sodium chloride) being recycled by winds off the playas.[3]

Besides salt, large quantities of soil also move off the playas. Mud dunes are formed when salts crystallize on the playas after rain or snow and cement soil particles together into structures the size of salt grains. These salt-soil structures obey the aerodynamic rules of sand grains. They bump and bounce across the playa surface in a wind-driven pro-cess called saltation. Once these large particles are driven off the playas into areas where there is some vegetation, they are trapped, forming dunes that look like sand dunes. When it rains on the mud dunes, the salts dissolve, the clay particles are freed, and the dunes become mud mounds that make travel virtually impossible.

The vegetation of these vast old lake-bed environments is dominated by low, widely spaced shrubs. Most of these shrubs belong to the goosefoot family, Chenopodiaceae. One of them, winterfat, is among the most valu-able shrubs for wintering livestock in the salt deserts. This white, wooly shrub has always suffered from too many synonyms. It is widely known as white sage. This name is unfortunate because it causes confusion with *Artemisia cana*, which is also known as white or silver sagebrush. The Idaho butcher who accidentally discovered the grazing value of winterfat started a trend. Shrubs that cattle readily graze by preference and that supply an adequate diet are rare. After the first hard frost in autumn, live-stock eagerly eat winterfat's prolific and highly digestible seeds. How-ever, winterfat is not tolerant to excessive grazing.[4]

Other members of the goosefoot family, especially the saltbushes, characterize the salt desert environment as well. The saltbush genus embraces about 150 species. About 60 of these species occur in the western United States. The Great Basin, with 32 species, appears to be the center of distribution for the western species. The new habitats that opened up in western North America after the recession of Lakes Bonneville and Lahontan allowed a host of new species of saltbush to develop.[5]

Of the valuable browse species, fourwing saltbush is one of the most important. Fourwing saltbush is a freely branched shrub, six to ten feet tall, with grayish white stems and leaves and rigid twigs. The name fourwing derives from the shape of the fruits. When the fruits are mature, livestock eagerly devour the seed crop. Fourwing saltbush is found over vast areas of the salt deserts, but it is generally mixed with other species and is rarely dominant. An exception is areas of sand dunes where the genetically different gigas (or giant) form of fourwing saltbush is found. Fourwing saltbush also occurs in patches around salt flats that are subjected to seasonal flooding.[6]

Shadscale is the dominant species over millions of acres of rangeland within the sagebrush/grasslands. In contrast to winterfat and fourwing saltbush, shadscale is not a preferred browse species. After shadscale plants lose their fruits in the fall, the branches of the short inflorescences on which the fruits were borne become rigid and then spinelike. These naked spiny branches persist for several years and provide considerable protection for the shrub against browsing animals. The leaves and fruits of shadscale drop off in autumn and are collected in soil surface depressions or form little wind drifts behind the bushes. These piles of leaves and seeds are the first thing livestock eat when turned onto salt desert winter ranges.[7]

Shadscale is the dominant species on many arid soils that are not exceptionally salty. The shadscale plants do well because they are naturally spaced at wide intervals in balance with the limited potential of the sites. Once the shrubs establish, soil particles accumulate beneath their branches and increase the area available for root growth.[8] The soil surfaces between the shadscale plants are stabilized with desert pavement, a network of small rocks left on the surface when fine salt particles are eroded by winds and rain. The rocks are covered with desert varnish and provide the background colors that characterize the landscape. When the pavement stones are derived from black basalt, the interspaces are as dark as burned landscapes and provide a preview of the popular concept of hell.

The only sagebrush species that is abundant in the salt deserts is budsage, which prefers the salt desert environment and has the distinction of being a spring bloomer. In February, when the mountain ranges are still firmly locked in winter snows and the nights are bitterly cold in the desert valleys, budsage begins to put out its vivid green leaves. Budsage occasionally coexists in winterfat communities as an understory shrub. During most of the year the winterfat communities are the wooly gray-green color of winterfat herbage. In the early spring, however, the startling green of the budsage leaves followed by their clustered yellow flowers makes the desert traveler wonder where in the world this new species came from. At the first hint of warm temperatures, the budsage leaves turn brown and wilt. The leafless branches are virtually invisible until the next spring. The browse of budsage is valuable food for cattle. Its early growth, when other forage is dormant and dry, is especially attractive to wintering livestock. In the 1890s budsage was considered one of the most important forage species for wintering livestock in the Red Desert of Wyoming.[9]

No fences limited cattle to the salt deserts during the winter. The cattle were free roaming, and between storms they drifted back into the sagebrush zone to graze and browse on preferred shrubs such as bitterbrush and cliffrose. The deserts were safe from deep and prolonged snow. Once the cattle were forced into the salt deserts by winter storms, their movements were restricted by water distribution. There is virtually no flowing surface water in salt deserts, so the cattle had to remain near springs.[10]

To understand the wintering of cattle on the salt desert, one should differentiate between free-roaming cattle and cattle forced to use salt desert areas. When the cattle were free to drop down into the desert valleys when winter storms struck the sagebrush, access to the salt desert served as a safety valve. In contrast, when cattle were driven out onto the salt desert and left at an isolated spring to forage for themselves, with no possibility to go elsewhere, the salt desert could become a deathtrap. The salt deserts were generally free from winter precipitation, but not from winter cold. Cattle can withstand a great deal of cold if their rumens are full. Cattle placed on a starvation diet, however, are susceptible to extreme cold, and the salt deserts are very cold deserts.

Wintering cattle in the salt deserts was like trying to fit a square peg into a round hole. The predominant forage species in the salt deserts are shrubs, and cattle are essentially grazers of grasses. Most of the valuable browse species, except for winterfat, are spiny plants protected from wholesale grazing. Cattle have a high daily water requirement and can travel only a limited distance from watering points. Sheep, in contrast, are selective browsers with a much lower daily water requirement. In fact, sheep can graze on salt desert ranges and depend on skiffs of snow for water.[11] This discovery ranks as one of the most important innovations in livestock husbandry in the nineteenth century.

The only forage grass that was abundant on nonmeadow areas in the salt deserts was Indian ricegrass (*Oryzopsis hymenoides*). Accounts of nineteenth-century grazing in the salt deserts often refer to this species as sandgrass. On winter range areas, Indian ricegrass is highly preferred by all classes of livestock and is rated as good to very good forage for cattle. The plants produce an abundance of herbage that cures well on the stalk and is very nutritious. The plump seeds are also high in food value and are sought by grazing animals. Stockmen of the nineteenth century had a high regard for this grass as winter forage. They called it a "warm feed" because it sustained livestock during severe winter weather.[12]

Water was vital if cattle were to use Indian ricegrass winter ranges.

Sheep can go without water for long periods when the feed is succulent. Horses on the range frequently water only at three-day intervals. Cattle need water every day to perform efficiently. When moisture falls on the winter ranges, small puddles of water form on depressions in the playas. Range-wise cows quickly extend their grazing area and work around these puddles until the last drop of foul, alkaline water is consumed or evaporates.

Indian ricegrass is first grazed near water holes. Gradually the cattle enlarge a circle around the watering points until all the herbage within a four-mile radius has been consumed. In the depths of the Carson Desert this type of winter grazing is still practiced. At distances greater than four miles from water the cows become much more selective and consume less and less of the coarser portions of the ricegrass plants. At ten miles from water, the most vigorous cows make only quick passes through the Indian ricegrass stands consuming the seeds that persist on the multibranched seed stalks.

If moving down into the depths of the desert valleys to escape the cold and snow of winter in the sagebrush was akin to dropping into the ranges of hell, then ascending the mountains to summer ranges must have been the cowboy's version of a journey to heaven. I. C. Russell described May Day as an event celebrated during August in the mountains that rim the western edge of the Great Basin.[13] On the western edge of the Intermountain area it was possible to escape the sagebrush environment by moving to the Sierra Nevada, Cascade, Steens, or Warner Mountains. Across the northern portion of the Intermountain area, the mountains of northeastern Oregon and the Idaho batholith provided summer range. On the east, the abrupt boundary of the Wasatch Front and the outlying ranges of the Rockies provided summer range.

The mountains that form the headwaters of the Humboldt Range in the northeast, the Toiyabe and Shoshone Ranges at the head of the Reese River to the south, and the Santa Rosa Range at the head of the Little

Humboldt have sufficient highland areas to support extensive areas of summer range and even have alpine areas. The many lower mountain ranges that subdivide the Great Basin support better-than-average rangelands, depending on the elevation and latitude. The mountains that rim the Intermountain area support mixed coniferous forests with many species of evergreen trees. In the Great Basin even the high mountains lack extensive coniferous forests. From the Truckee River in western Nevada to southeastern Idaho and southward through Nevada and Utah, the lower slopes of the mountain ranges have open woodlands of single-leaf pinyon and Utah juniper commonly called pinyon/juniper woodlands. Both species are multitrunked and bushy.

Pinyon/juniper woodlands are not lush, shaded forests. With their irregularly shaped crowns and thickened bases, pinyons hunker down to the harsh, rocky soils. The trees give the appearance of having had to fight for survival in a generally treeless environment. The needles of single-leaf pinyon are sharp and flecked with resin droplets. An unwary traveler who brushes against a pinyon receives a sharp prick and a lasting coat of pitch as a reward.

Pinyon seeds were an important part of the diet of the Indians native to the central Great Basin. The seeds were eaten raw, roasted, or cooked, and were preserved in the form of pinole, a finely ground flour. Pinyon resin was used as chewing gum as well as for mending, cementing, and waterproofing. In the fall when the juniper berries are ripe, the pinyon/juniper woodlands are alive with the harsh cries of pinyon jays, which feed on the berries.

Both pinyons and junipers, especially seedlings and young plants, are susceptible to wildfires, although the habitat determines this to some degree. An aerial photograph of the Palisade portion of the 1964 fire in Elko County showed that the pinyon/juniper woodlands in the valleys and on the north slopes were burned over and killed. Those growing on steep, south-facing slopes were unburned because there was insufficient understory vegetation to carry the fire from tree to tree. All age classes of trees, from century-old patriarchs to seedlings, were present. The trees that burned were relatively even-aged stands. These trees had invaded the sagebrush/grasslands, moving out from the true pinyon/juniper stands growing on fire-safe sites. This process has apparently been going on for thousands of years since the close of the Pleistocene.[14]

Above the pinyon/juniper woodlands on the interior mountain ranges are extensive areas of mountain brush that form summer range for cattle and mule deer. They appear to be a continuation of the sagebrush/grasslands. Although big sagebrush is a major component of these communities, it is usually represented by one subspecies, mountain big sagebrush, which is more highly preferred by browsers. Besides mountain big sagebrush these communities contain the valuable browse species bitterbrush, curlleaf mountain mahogany, and species of snowberry. The mountain brush communities support productive grasslands dominated by bluebunch wheatgrass or needlegrass species and Idaho fescue on the north slopes.

Mountain meadows are an important part of the summer range environment. Fertile soils and a good soil moisture balance from a high water table contribute to a productive potential much greater than that on adjacent sagebrush ranges. These meadows range in area from a few square feet to several hundred acres and are dominated by tall grasses such as Nevada bluegrass and meadow barley. Under the tall grasses is a tuff sod of sedges, wiregrass, and mat muhly. Forbs include Rocky Mountain iris and pale agoseris.[15] These meadows are often called "stringer meadows" because they form narrow green strips along watercourses through a generally gray, shrub-dominated environment. Meadows are important habitat for sage grouse, the "sage chicken" characteristic of these environments.[16]

On mountain ranges that exceed nine thousand feet in elevation, scattered stands of five-needle pines—the ancient bristlecone, white bark pine, and limber pine—occur above the mountain brush. Aspen groves are also characteristic of summer ranges. Quaking aspen is one of the few deciduous trees widely distributed in the uplands of the sagebrush/ grasslands. On major mountain areas, extensive aspen groves, composed of both trees and shrubs, provide browse and shelter. In the shade of the aspen patches and near perennial snowbanks grow tall larkspur plants. These violet blue flowers are deceptively beautiful. Larkspur plants contain poisons that are especially deadly to cattle. They claim their highest toll early in the season because larkspurs start to grow before other forage species. Sheep are less susceptible to the poison and are seldom poisoned in the field. Tall larkspur poisoning limits the use of some summer ranges by cattle.[17]

Under pristine conditions the high mountains of the Intermountain area supported desert bighorn sheep. Three subspecies of mountain sheep are thought to have occurred in Nevada—the desert bighorn, the California bighorn, and the Rocky Mountain bighorn. For feeding routes, bedding grounds, and all-around living quarters, the bighorn prefers the roughest and most precipitous country on or near the mountaintops. When forced to travel in lower country, the sheep gravitate to high spots that allow them to see the surrounding country. Bighorns are browsers and grazers of forbs and grasses. Each band has a regular feeding route that usually terminates at a bedding ground. Snow and winter cold drive the bighorns downslope to desert ranges. For drinking, where a choice exists, the sheep select springs or water holes in the highest, most inaccessible parts of the mountains.[18]

The elk (or wapiti), the largest member of the deer family on sagebrush ranges, was almost absent from the Intermountain ranges under pristine conditions. In the Pacific coast region, elk occupied the Coast and Cascade Ranges from Vancouver Island southward through Washington, Oregon, and into southern California. Curiously, most of Arizona and Nevada and parts of Utah, Oregon, and Washington appear not to have had elk populations.

The only historical mention of elk in Nevada comes from Captain J. H. Simpson's journal of his exploration for a wagon route from Camp

Floyd, Utah, to Genoa, Nevada. Simpson's entry for July 20, 1859, states: "An elk was seen for the first time yesterday in Stevenson's Canyon, and one today in Red Canyon, also a mountain sheep for the first time." Red Canyon is in the Snake Mountains just north of Wheeler Peak, and Stevenson's Canyon is west of the Schell Creek Range, both in eastern White Pine County.[19]

Mule deer populations in the Intermountain area are quite variable. In much of the Great Basin, the quantity and quality of summer range often determine the herd size. In the more arid portions of the Intermountain area, only the highest mountains provide suitable habitat for mule deer. Thus, the mountains of northeastern Nevada provided excellent mule deer habitat. Many settlers looked on the big game as a source of winter meat. For example, Utah residents from Brigham City made hunting trips to the mountains of south-central Idaho to kill mule deer. In 1883 T. D. Calahan and John Strude returned from the mountains of Owyhee County with forty-five mule deer carcasses.[20]

An almost universal characteristic of summer ranges in the sagebrush/grasslands is that they are steep and rugged. There are very few level uplands. Cattle graze level land best and tend to congregate on level areas such as stringer meadows. Cattle can traverse amazingly steep land, but the willingness, not the ability, of an animal to get into steep areas to graze is what is important. Sheep are better adapted to grazing steep topography. Their smaller size enables them to negotiate steep areas more easily. And since they are ordinarily under the control of a herder, they can be forced to graze steep slopes.

The winter escape to the desert and summer escape to the mountains enhanced the potential of sagebrush/grasslands ranges to support cattle. Successful ranch management required a balance between these resources. A given amount of sagebrush/grasslands had to be balanced by winter and summer rangelands. Later, the requirement of hay lands was added to this balance.

Satellite winter and summer ranges have some basic weaknesses that make them more suitable for raising sheep than cattle in the cold desert. First, the forage on winter ranges is mostly browse species, and use of this browse is limited by lack of water. Sheep are better adapted to wintering on the salt deserts because they are more efficient users of browse and require less water than cattle. Second, mountain ranges, while providing a desirable escape from the summer drought in the sagebrush, can also be rugged, steep, and infested with tall larkspur. Sheep are better adapted for grazing on steep subalpine slopes than cattle and usually are not affected by tall larkspur poisoning. In the 1870s and 1880s these appeared to be small and unimportant points, but they left the door open for competition between cattle and sheep ranchers.

CHAPTER 4

Texas Cattle and Cattlemen

To many, "cowboy" equals Texas. Millions of words in story, song, and history have been written about this connection. Although this book deals with the Intermountain area, the history of Texas cattle and cattlemen must be explored to trace the roots of and give perspective to cattle in the cold desert.

In the late seventeenth and early eighteenth centuries, the French and later the Spanish made abortive attempts to establish colonies in the territory that would become Texas. Spain established a system of missions in south Texas but met savage resistance from the Indians. After several false starts, the Spanish finally achieved a measure of control. But each time their missions were abandoned, the cattle they had brought with them were left to run wild in the brush lands. After the Mexican Revolution of 1821, the Texas territory passed to Mexico.[1]

American colonists were moving into the Mexican territory of Texas throughout the early 1800s. Many came from the South and knew how to ride and shoot, but had no concept of the Spanish system of open-ranging livestock. When Stephen F. Austin proposed a code of law to the Mexican government to govern Texas, the Mexicans added two articles they felt necessary. One article dealt with the registration of brands; the other pertained to the disposition of strayed stock.[2]

The contributions of the Spanish to the western American livestock industry have been stressed repeatedly, especially by Walter Prescott Webb and J. Frank Dobie. More recently, Terry G. Jordan has emphasized the role of American colonists from the South in the roots of the Texas livestock industry.[3] Both schools of thought have merit, and both cultures made contributions. The role of livestock in the early settlement of the South was long overlooked, and Jordan is to be commended for bringing recognition to the southern stock raisers. But in their eagerness to upset tradition, some supporters of Jordan's ideas substantially discount the influence of the Spanish on the evolution of ranching in the West.

The literature of the southern livestock industry carries numerous references to cow pens and herding. The southern environment permitted year-round grazing and encouraged winter grazing. But on the ranges of the West there would be no pens for working cattle and no herding in the northern European style. The western technique of mounted men holding cattle on an open piece of ground (rodeo ground) while other mounted men sorted them was completely foreign to southern livestock production.

After the Civil War there was a growing demand for Texas beef all over the United States. The northern states had prospered during the war, and their market needs were building. The huge crews building the new transcontinental railroad system required food at many points. The vast area acquired by the United States through the Louisiana Purchase and Mexican War was developing. Much of this land was suitable for stock production, and there was an immediate demand for red meat in the far West mining boomtowns.[4]

Texas Longhorns are descendants of cattle brought to the New World from Spain. The Retinto is doubtless the major breed of Spanish cattle represented in the Longhorn. But the remnants of relatively pure Spanish bloodlines found in isolated regions of the Western Hemisphere include several regional breeds of Old Spain.[5]

A composite of Frank Dobie's descriptions provides an accurate portrait of these quintessential cattle of the Old West.

> Longhorns were tucked up in the flanks with high shoulders so thin they sometimes cut hailstones. Their ribs were flat and their length frequently was so extended that the back swayed. Viewed from the side, its frame would fool one into overestimating its weight. A rear view showed cat hams, narrow hips, and a ridgepole backbone. The tails of longhorns often dragged the ground despite their racehorse legs. The colors of longhorns illustrate what they were, a free breeding, outcrossing population of free roaming animals.[6]

Longhorns were slow to develop, not reaching their maximum weight until they were eight to ten years old. Yearling Longhorns were lucky to reach 300–350 pounds. Three-year-old steers might reach 500 pounds, and five-year-old steers 750–800 pounds. Steers from four to eight years old averaged 800 pounds. Animals ten years old might reach 1,000 pounds, and rare exceptions reached 1,600 pounds. This was an extremely slow growth rate compared with English breeds.[7]

Longhorns were slow converters of feed to beef and poor producers of the heavier cuts. Because they were long and lean and lacked a paunch, they dressed out surprisingly well. Their beef was inferior and had a high percentage of bone compared to muscle. By an Indiana man's estimate, "one could salt in its horns all the roasting beef an average Texas steer could carry."[8] Although the Longhorn was generally a poor-quality animal, spectators at any American cattle market in the 1860s saw more bad than good specimens of cattle.[9]

The conditions under which the Texas Longhorns were produced contributed to their poor quality. They received no supplemental feed, summer or winter. They roamed the grasslands and the woods for forage. The average winter death loss was 20 percent. Aside from the quality of the beef it produced, the inherent cruelty of the system was repugnant to many herdsmen. James MacDonald reported, "The prairies here and there are strewn with whitened skeletons; only an acclimatized Texan could contemplate with equanimity the fate of these unfortunate famished animals. At one side station more than fifty two-bushel bags full of bones were lying ready for transport."[10]

One thing could be said in behalf of the Longhorns—there were a lot of them. In the 1870s MacDonald calculated that America required eight brood cows per ten people to meet the annual consumption requirement for beef. In Texas there were more than nine cows per person! The 1830 census estimated 100,000 head of cattle in Texas.[11] The 1850 census estimated 330,000, and that of 1860 estimated 3.5 million head.[12]

From the Texas revolution until the Civil War, cattle ran wild in Texas, multiplying rapidly. Sporadic attempts were made to transport them to the New Orleans market, to California after the Gold Rush, and even to the North, but nothing was standardized. By 1837 cowboys had begun gathering herds of 300–1,000 wild, unbranded cattle in the Nueces Valley and driving them to Texas cities to sell. In 1842, drives to New Orleans began and there was a drive to Missouri. In 1846, Edward Piper drove cattle to Ohio for fattening. From 1846 to 1861 the drives increased. In 1850, drives went to California; in 1856 to Chicago. Until the Civil War there was continual, though irregular, movement of cattle out of Texas.[13]

In 1865, cattle in Texas could be bought for three to four dollars per head but found few buyers. The same cattle in the northern markets would have brought thirty to fifty dollars. Many Texans made vigorous efforts to connect the four-dollar cow to the forty-dollar market. In fifteen years, five million cows were delivered to the northern market twelve to fifteen hundred miles away.[14] Sedalia, Missouri, was the closest railroad shipping point, but southwestern Kansas, southern Missouri, and northern Arkansas were hostile to Texas cattle—a combination of fear that Longhorns transmitted Texas fever, ill feeling from the

Civil War, and frontier lawlessness. On the whole, the season of 1866 was disastrous for Texans who tried to drive cattle to Missouri.[15]

J. C. McCoy was the first man to see the desirability of a point of contact for eastern buyers and Texas drovers. His plan was "to establish at some accessible point a depot or market to which a Texas drover could bring his stock unmolested, and there, failing to find a buyer, go upon the public highways to any market he wished." It was to establish such a market, where the southern drover and the northern buyer could meet upon equal footing undisturbed by mobs or thieves, that McCoy established the original "cow town" of the West—Abilene, Kansas.[16]

At times there was no market. Surplus cattle were held "on the prairie" or on ranches to be fattened. Thus, the cattle kingdom spread outward from Texas, utilizing the grasslands of the plains. Texas furnished the base stock, the supply, and the method of handling cattle on horseback. The plains offered free grass. According to Webb's classic assessment, "From these conditions and from these elements emerged the range and ranch cattle industry, perhaps the most unique and instinctive institution that America produced."[17]

Marketing of Texas cattle through westward-spreading Kansas towns continued through 1873. Until 1870, herds sent to Abilene and other railroads sold in a steady or rising market. Prices were particularly good in 1870, and the movement from Texas in 1871 was the greatest in history—700,000 head went to Kansas alone. But late in 1871 market conditions changed and drovers met a reversal. Half the cattle brought from Texas remained unsold and were wintered at a loss on the Kansas prairies. On September 18, 1873, the New York banking firm of Jay Cook and Company closed its doors, precipitating the first financial panic known to the range cattlemen. Thousands of Texas cattle were tanked for their horns and tallow, with drovers receiving from one to one and a half cents per pound.

J. Frank Dobie writes, "After the Civil War all it took to become a cattleman in Texas was a rope, the nerve to use it, and a branding iron."[18]

One of the more prominent stock drovers was Colonel D. H. Snyder, born in Mississippi on September 5, 1833. Moving to Texas in 1854, he settled in Williamson County. He bought Spanish horses in south Texas and drove them to Missouri, where he traded them for draft horses. Building on his horse-trading experience, he delivered beef to the Confederate army. His herds forded the largest rivers, including the Mississippi. He had three water-trained steers at the head of his herd. When these leaders took to the water without hesitation, other members of the herd followed safely.[19]

Colonel Snyder and his brother, J. W. Snyder, made one of their first early postwar cattle drives to the West. They bought the cattle in Llano and Mason Counties, Texas, paying $1.50 per head for yearlings, $2.60 for two-year-olds, $4.00 for cows and three-year-olds, and $7.00 for beef steers. They bought on credit with notes payable in gold coins. The cattle were driven across west Texas on the trail pioneered by John Chisum to Fort Union, New Mexico, where the brothers sold some of the cattle for $35.00 a head. They continued to Trinidad, Colorado, and sold the remainder to Charles Goodnight at $7.00 per head without requiring tally.[20]

In 1866 the Snyder brothers drove stock from Llano County to Abilene, Kansas. After losing 140 head to Indians in the Indian Territory, they filed a claim against the federal government and were eventually paid for the loss. Colonel Snyder abolished the practice of slaughtering calves during long drives. On one of his initial northwest drives, after a day spent in bloody slaughter of calves of the herd, he issued orders that the slaughter be stopped. His herd arrived at its destination in good condition, and he realized a fair profit from the calves, which weathered the waterless waste as well as the mature stock.

Colonel Snyder ran a taut ship, and possibly his greatest contribution to ranching was the training he gave young Texans in organization and leadership. The Snyder brothers had three rules of conduct for their cowboys: (1) you could not drink whiskey while in their employment; (2)

you could not play cards and gamble while working for them; and (3) you could not curse and swear in their camps or in their presence. One of their brightest young cowboys, John Sparks, may have found some of those rules difficult to follow.[21]

The Snyder brothers changed tactics in 1870 by driving a large herd to Schuyler, Nebraska, seventy-six miles west of Omaha. This was the first herd to cross the Kansas-Pacific Railroad and continue to the Union Pacific. In 1871 they delivered four large herds to the Platte River valley to establish their northwest headquarters at Cheyenne, Wyoming. The cattlemen of the plains were continually on the lookout for new range and ranching areas.

In 1872 the Snyder brothers delivered a large herd to a Texan in the far northwest corner of Utah on Goose Creek. The Texan was John Tinnin, an ex-Confederate "Colonel" doing business in Utah as Ingram Company. Colonel Snyder selected John Sparks to lead this drive across the mountains to the Great Basin because Sparks had visited the area in 1868 while working for Colonel John J. Meyers. John Sparks would spend most of a decade east of the Rockies before he returned to the sagebrush/grasslands, but these early trips introduced him to the area. Drovers called Meyers the "Father of Israel" because he drove cattle into the wilderness. Records are not available, but most likely the Ingram Company was a livestock commission company and the cattle delivered to Goose Creek were for Jasper Harrell.

The Snyder brothers made their biggest drive in 1873, only to find on reaching Wyoming that financial panic had destroyed the market. They borrowed money at 36 percent per annum to hold the herd through the winter. They managed to sell part of the herd to a firm in Salt Lake City, so the brothers shod work oxen, bought fresh horses, and took the cattle across the mountains. When they reached their destination, Congress had demonetized silver, the banks were closed, and their Utah buyers could not fulfill the contract. Turning north to the Fort Hall Indian Reservation, Colonel Snyder entered bids to supply beef for the reservation.

He entered bids for each Snyder brother, half-brother, and brother-in-law. The army accepted one of the bids, and the brothers were in business on the sagebrush/grasslands of the Intermountain area. They delivered twenty-five beefs a week to the reservation, with the rest of their cattle ranging as far west as Rock Creek on the Snake River Plains.

The Snyder brothers contracted in 1877 to deliver cattle to J. W. Iliff, a cattle baron of Colorado, and delivered twenty-eight thousand head of Texas stock. After Iliff's death, they became managers and administrators of his estate. They continued to supply Texas cattle to the northwest until 1885 and were some of the first, last, and largest volume stock drovers. The Snyder brothers were instrumental in supplying stocker cattle to establish the range livestock industry in northeastern Colorado, western Nebraska, Utah, and Wyoming. In 1885 Colonel Snyder was offered one million dollars for his northwest interests. He refused, only to lose most of his holdings in the hard winter of 1886 and the depressed markets that followed. After that, the Snyder brothers returned to Texas and became ranchers. The big days of the overland trail drover were past.

Other Texans did more than come and go in the Intermountain region; they transferred their cattle, their ranching operations, and their expertise to the sagebrush/grasslands. The Texan who most influenced livestock production in the area during the nineteenth century was probably John Sparks. John Clay, a Scottish-born livestock man who was widely respected in the American West, thought that Sparks's Texas background had given him a wonderful knowledge of the cow business. Clay described Sparks as "tall, straight as a pine tree, with a clear flashing eye that seemed to look right through you. While somewhat deliberate in his movements, he was full of energy."[22] Sparks stood out in any crowd. He sported English corduroy pants, western riding boots, a flask of whiskey, and a six-shot Colt revolver. In the winter, he topped this regalia with a coonskin coat. Later in life, as a governor of Nevada, he was popularly known as "Honest John" Sparks.[23]

What factor in Sparks's background enabled him to develop a livestock empire? The roots of his family originated in England. The American branch of the family arrived in Maryland in the late 1600s, well before the Revolutionary War. John Sparks's grandfather, Millington Sparks, was a moderately successful planter in Maryland. John's father, Samuel Wyatt Sparks, apparently decided to move the family's endeavors to the promising new South.

John Sparks was born in Winston County, Mississippi, on August 30, 1843, the seventh of ten children.[24] His mother, Sarah, was a Deal from South Carolina. Later in his life John Sparks was described as having the manners of an Old South aristocrat. His was one of the early "new lands" families moving across the South with the forward edge of developing agriculture. The Sparkses were not plantation agriculturalists in the southern pattern of cotton and industrial crops. They developed new land, either brought it to production or depleted it, sold their improvements, and moved on.[25] The family moved from Mississippi to Fountain Hill in Ashley County, Arkansas. The 1850 Arkansas census rolls list Samuel Sparks as a farmer owning real estate valued at twelve hundred dollars.[26]

The family continued its westward trek, moving to Lampasas, Texas, and settling about four miles east of town on Burleson Creek. This area of Texas borders on the black-land prairie soils that eventually were developed for cotton production. The census of 1860 indicates that Samuel Sparks owned real estate valued at six thousand dollars with improvements valued at twenty-three hundred dollars. The family had done fairly well in the 1850s.[27]

When the Sparks family moved to Texas, John was fourteen years old. He later said that he started in the cattle business at that point. He became proficient in riding, roping wild cattle in the brush, and following the Dobie axiom: "a horse, saddle, rope, branding iron, and the guts to use them." He stayed with the family ranch until the outbreak of the Civil War when he was eighteen.

In view of the manpower needs of the Confederacy compared with those of the more densely populated North, Sparks was destined to become a soldier. Texas was the only southern state that was also a frontier state. Texans felt mixed emotions at the outbreak of the war. The former republic strongly supported the concept of states' rights. But most settlers on the Texas frontier did not have slaves, nor did they necessarily want to fight for the institution of slavery. John Sparks may have fallen into this latter group. He found a compromise that allowed him to avoid combat against the North, but the option he selected was equally dangerous.

From the time the Spanish had first tried to establish missions in Texas, the Comanche had offered stubborn resistance. Anglo settlers had pushed them across Texas in many bloody encounters. When federal troops withdrew at the start of the Civil War, the Indians were quick to take advantage. To cope, Governor Lubbock of Texas established the Frontier Regiments of Texas cavalry under the Texas Rangers as an alternative to Confederate army service. Many of the young Texans who later became leaders in the range livestock industry chose this form of military service. On January 29, 1862, just over a thousand Indian fighters were organized in nine companies of Texas Rangers under the command of Colonel James M. Norris. By spring they occupied eighteen outposts from Gainesville on the Red River to Fort Duncan on the Rio Grande.[28]

John Sparks enlisted as a private in Company 6 on March 5, 1862, and was mustered into service at Keens Ranch. Records indicate that Sparks received $0.40 per day for each of the three horses he furnished and $5.47 for the arms, plus a salary of $12.00 per month. Each man in the company carried two revolvers in his belt, two in saddle holsters, and a rifle in a scabbard. They trained by holding shooting matches and riding tournaments, with plenty of whiskey, women, and song. This was a rugged frontier action school for the young John Sparks.

In later years Sparks was often referred to as "Captain" John Sparks based on his Civil War service. The limited records available indicate that he went into the service as a private and was discharged at the same rank, serving from age nineteen to twenty-three. Considering the number of Texans who became "colonels" after the war, "captain" was a modest promotion.[29]

Sparks started back in the cattle business in Texas as soon as the Civil War ended. By 1868 he was driving and delivering Texas cattle to Wyoming. As mentioned above, he got his first look at the Great Basin on a drive to northwest Utah while working for Colonel John Meyers. Sparks described himself in those years as "employed by others and receiving a very small salary." He once delivered a large band of Longhorns from Texas to Virginia. He drove them first to Memphis, Tennessee, a trip of nearly five hundred miles. There they were loaded on cars of the Memphis and Charleston Railroad and shipped to Alexandria, Virginia. The steers were unloaded at Bristol, Tennessee, for forage and water. When Sparks stopped at Bristol on the return trip, he was upset to find local cattle dying from Texas fever contracted from his stock's grazing area.[30]

John went to Wyoming after his trip to the East and remained there for two and a half years. In 1872 he returned to Texas and married Rachel Knight, the daughter of Dr. D. J. Knight of Georgetown, Texas. During the same year he and his brother Tom drove cattle to Wyoming and sold them at a good profit.[31]

Georgetown, Texas, was John Sparks's home and operating base for many years. It was also the Texas base for the Snyder brothers. In later years, one of John Sparks's cowboys, Tex Willis, wrote an interesting letter to Fred W. Sparks, a professor of mathematics at Texas Tech University. The cowboy and former wagon foreman wrote that

John Sparks did the same amount of work Charlie Goodnight did, except John made no noise. G. W. Littlefield, Ike Rogers, Ed Anderson, Ike T. Pryor all worked on the trail to Nevada with "old John," as they called him and swore he was the greatest cowman they ever knew or worked for, but knew little about him. John was a top member of Texas Frontier Battalion in 1861, along with Charlie

Goodnight. And in 1868, John Sparks got to California with his first 4,000 Texas cattle, and was paid off in an enormous amount of gold money for that day in time. Old John said simply, "You have to trail cattle where people have gold money to pay for them."

In the early 1870s John and his older brother Tom moved "four gigantic herds of cattle." Apparently, all the drives were profitable. According to historian J. H. Triggs, Tom Sparks "was influenced by the rich Fort Hall bottomland to move to Idaho with his herd of 14,000." Tom's operation was later wiped out in the big freeze of 1889–90.[32]

In 1873 John Sparks bought a large herd in Texas and drove the cattle to Wyoming. He established a ranch in the Chugwater River valley and moved his wife and family to Cheyenne. The Chugwater River valley is about seventy miles long and three miles wide and covers an area of about 135,500 acres. The greater part of its length is bordered by high, rocky walls, and its fertile bottomlands are quite low and level. The water supply is sufficient for irrigation of the natural meadows.[33] Hundreds of work cattle had been watered in the Chugwater River valley as early as 1852 by Seth E. Ward, a trader.[34]

Ranching had developed rapidly in Wyoming during the late 1860s. Nelson Story trailed Texas cattle to Wyoming in 1866; the Union Pacific railroad reached Cheyenne in 1867; and by the fall of 1868, 300,000 Texas Longhorns had reached Wyoming.[35] When John Sparks arrived, in the early 1870s, Wyoming was swiftly changing. Settlers and their cattle were threatening the existence of the Native Americans there and the American bison on which the Indians relied. The government seemed to support the rapid extinction of the bison to end the independent existence of the Plains Indians.[36]

Wyoming presented a more arid and much colder environment for the range livestock industry than Texas. The average annual precipitation at Cheyenne was roughly sixteen inches, versus thirty inches in Williamson County, Texas. Both areas received 70 percent of the precipitation dur-

ing the growing season.[37] But aside from money used to purchase cattle,
investment in these early Wyoming ranches was slight. A dugout cut into
a hillside near a creek served as headquarters, with a similar dugout
nearby for the horses.[38]

In 1873 the Chugwater River valley was stocked with 4,100 head of
cattle; 2,700 of them belonged to John Sparks. One might expect the
young, recently married cowman to settle down and develop his ranch.
But in 1874 Sparks sold his cattle and land claim to the Swan brothers.
This was the first purchase by A. H. and Thomas Swan, who eventually
assembled one of the largest land and cattle firms north of Texas. The
purchase price was thirty-five thousand dollars; Sparks later told H. H.
Bancroft that it was "the largest sale ever made up to this time in the
West." The Swan brothers paid fifteen thousand dollars as a down payment. A chattel mortgage remained on the property until the balance was
paid, an unusual arrangement at the time. The amount of land involved
in the transaction is unknown. Apparently, no deeded acres were involved. Laramie County's General Index to Deeds and Mortgages for
1867–83 shows no record of the transaction. Federal Land Office records
in Cheyenne show no listing for John Sparks. Sparks may have established a preemption claim on the Chugwater Valley, but he never filed it.[39]

Sparks joined the Wyoming Stock Growers Association (WSGA) on February 23, 1874, at its second meeting.[40] In a single decade this organization grew from 10 members with 20,000 head of cattle valued at
$350,000 to 435 members representing 2 million head of cattle worth
$100 million. The WSGA provided a means of organizing community
roundups on the vast public lands, of protecting large stock owners from
theft by registering brands and hiring brand inspectors, and of developing organized political power in the territorial government.[41]

Sparks later recalled that in 1875 he "purchased a herd in Colorado,
drove them to Wyoming, and established a ranch on the North Platte
River. This was the frontier ranch of that section of the country at the
time. The Sioux Indians were so troublesome as to make it absolutely

unsafe to go further into the interior."[42] Within a year, in 1876, Sparks sold this property to Sturgis and Carre of Cheyenne. The agreement of sale in the Wyoming State Archives lists 3,000 head of cattle, plus ranch and horses. Then Sparks immediately purchased 5,000 head from the Swan brothers and established a new ranch on the North Platte River near Fort Fulterman. Sparks sold 2,700 head in 1874, and then bought 5,000 back in 1875 from the Swan brothers. He was pyramiding his capital by rapidly following the retreating Indian frontier and establishing ranches, which he just as rapidly sold for profit.

Sparks made frequent trips from Texas to Wyoming to look after his business interests. John and Bill Blocker delivered three thousand head of cattle to a Sparks ranch west of Cheyenne in 1877. In 1878 Sparks purchased a ranch, range, and a large herd of cattle on Lodgepole Creek, on property on the Wyoming and Nebraska territorial line. He sold it before the end of the decade to the Bay State Company. The Lodgepole Ranch became a very famous ranch during the cattle bonanza period of the 1880s.[43]

Sparks did not plow back all of his pyramiding capital into Wyoming ranches. Among other investments, he became a 50 percent partner with M. E. Steele in a bank in Georgetown, Texas, that was active in financing the developing cotton production industry in the Williamson County area. The railroad reached Georgetown in 1878, just as the area was coming to life after the long economic depression of Reconstruction. One of the major accounts of the Steele and Sparks Bank was that of D. H. and J. W. Snyder.[44]

Sparks also purchased ten thousand acres south of Taylor, Texas. There is no available record of the purchase price, but even at a minimum of one dollar per acre it was a substantial investment. He bought land from the Knight family and built a house on South Brushy Street that was regarded as one of the finer homes in Georgetown. He continued to maintain this residence even after he was elected governor of Nevada and often entertained the governor of Texas there.

Personal tragedy struck Sparks in February 1879 when his first wife, Rachel Knight Sparks, died at age twenty-six. They had two daughters; Maude was born in 1874, and Rachel was born in 1877 but died at age three. On January 25, 1880, John Sparks married Rachel's half-sister, Nancy Elnora Knight. They had four sons, with three surviving infancy. Deal was born in 1880, Benton in 1882, Charles in 1885, and Leland in 1887.[45]

John Sparks was phenomenally successful in the cattle business throughout the 1870s. He was able to utilize his experience in a unique situation. It was a seller's market, and John Sparks was selling his claims, range he did not own, primitive capital improvements, and low-cost cattle he had driven across the rugged trails and frontier miles from Texas. Eastern and foreign capital was begging for the opportunity to buy. Titled young men from Europe and Ivy League graduates were rushing to Wyoming to invest in ranching and to have the chance to associate with those wild young knights of the plains—the Texas cowboys.

Part II **The Land Acquired**

Thermal winds swept through the spreading alluvial fans, rippling the endless strands of wheatgrass, Indian ricegrass, and needlegrass. Meandering stream channels supported sod-covered meadows in the canyon bottoms. The silver gray ocean of sagebrush contained islands rich in potential for grazing animals. There were men of courage ready to risk fortunes in a grand experiment to determine if the cold deserts of the Great Basin could support a society based on extensive herds of cattle.

CHAPTER 5

Buy, Beg, Borrow, or Steal a Ranch

When John Sparks and John Tinnin bought an empire from Jasper Harrell, in essence they bought Harrell's claim to specific meadowlands along streams. No law specified the right to use the rangeland; that was partially dependent on Sparks's and Tinnin's ability to defend its exclusive use.

A post–Civil War entrepreneur moving west to begin development of a stock ranch generally lacked sufficient capital to acquire outright ownership of land and livestock. The usual approach was to buy cattle cheap and use range free of charge. The vast bulk of the western states was owned by the federal government. Exceptions were the land grants given to railroads to facilitate construction and grants given to states when they entered the union. The government also granted land to be sold for the support of public schools, universities, and other public works. Both railroads and states offered land received in such grants for sale, under a variety of options.

The prospective rancher had ten options for obtaining land from the public domain during the late nineteenth century in the far western United States:

1. He could buy land offered at public auction by the General Land Office.
2. He could purchase land previously offered for sale at auction but not sold.
3. He could file a preemption claim.
4. He could file for a homestead.
5. He could file for land under the Timber Culture Act.
6. He could purchase land under the Timber and Stone Act.
7. He could file for land under the Desert Land Act.
8. He could file for land under the Mining Law.
9. He could file for land under the Coal Land Law.
10. He could obtain land by military-bounty land warrant.

The options ranged from common practices to obscure processes.[1]

The first two options dealt with distribution of land through public auction. Public lands were opened to auction following a proclamation by the president of the United States or notice by the General Land Office. This was the usual procedure as the frontier advanced to mid-America. Lands not sold at the auction could be purchased from the General Land Office at any time thereafter.

A third method was the preemption claim, or "squatter's right." Settlers could improve 160 acres of unappropriated public land, then later buy it at $1.25 per acre without competition. The preemption right was established by construction of a house and improvements. The settler had to file a declaration of intent to purchase within three months after settlement, or within three months of filing by the General Land Office on previously unsurveyed land. Preemption claims were often made far in advance of land surveys. Due to delays in obtaining surveys, many preemption claims were held for years without final payment. Payment could be made up to eighteen months after filing the declaration. Payment could be made in cash, a rare commodity on the frontier, or military-bounty land warrant, or agricultural college script.

A fourth method of obtaining public land was through homesteading. President Lincoln signed the Homestead Act on May 20, 1862, marking a reversal of earlier federal land policy. Before that date, the government adhered to the Hamiltonian notion that it was the federal government's responsibility to develop revenue and provide homes indirectly to settlers or pioneers. The Homestead Act marked the adoption of the idea that the interest of government might be best served by providing land to settlers and later receiving compensation from increased national prosperity and from property values that could be used as the basis for public revenues.[2]

Provisions of the Homestead Act permitted settlers to acquire 160 acres free of charge except for filing fees. The homesteader had to live on the claim for five years to receive title. The major differences between homesteading and preemption were the $1.25 per acre payment required under preemption and the five-year residency required on homesteads. An individual could combine the two methods to receive 320 acres of public land.

The Homestead Act reversed a long-standing policy of reducing the acreage offered at public land sales. Farmers were not interested in buying more land than a family could operate, and the eastern United States had seen a steady decline in the size of auction blocks, from 640 acres down to 80. Prospective farmers had no objection, however, to receiving 160 acres free of charge.[3] To a congressman from the East, the grant of 160 acres of public land to a settler seemed a big giveaway. Unfortunately, the environmental conditions of the West required much more than 160 acres to support a livestock operation.

Consider, for example, a hypothetical 160-acre ranch providing 1,250 pounds of annual herbage per acre. This herbage is produced from April through August with 80 percent usable by grazing animals. Each cow needs 12 acres per year (1,250 pounds herbage production 0.80 forage utilization = 1 Animal Unit Month [AUM] per acre 12 months = 12 acres). This gives a stocking capacity of thirteen cows per 160-acre homestead.

But a ranch cannot run just thirteen cows. The herd must have a bull and two replacement heifers; and steers are not marketable until they are three years of age. With a 100 percent calf crop, the herd would consist of four cows (4 AUMS), one bull (1.5 AUMS), two replacement heifers (2 AUMS), two yearling steers (1.5 AUMS), two two-year-old steers (2 AUMS), and two three-year-old steers (2 AUMS), for a total of thirteen animals. The two marketable steers would have a value of $20.00 each after three years. So, for the first three years of the five-year requirement, there would be a return of $13.33 per year. This hypothetical homestead collapses because there is insufficient year-round grazing. The 160-acre homestead was an economic and biological impossibility in the sagebrush/grasslands.

Smart homesteaders used their homestead claims to acquire title to land that could be irrigated. If 100 of the 160 acres claimed was irrigable, the homesteader might produce eighty tons of hay, sufficient to winter eighty brood cows. But in their annual migration from winter ranges through the foothills to the summer ranges on the mountains, the homesteaders' eighty-cow herd required some 1,960 AUMS in addition to the hay produced by irrigating. Part had to be in the desert and part in the mountains to satisfy seasonal forage requirements. Both areas contained extensive areas of salt flats, rock outcroppings, and other nongrazable areas. In seasonal migration, a homesteader's eighty-cow herd might have to graze over 10,000–15,000 acres for sufficient forage.

A fifth method of obtaining land from the public domain, the Timber Culture Act, was possibly the most bizarre of the options available to would-be landowners. Congress passed the act based on the assumption that trees planted on the Great Plains would increase rainfall and make dryland farming practical. This law granted 160 acres to the settler who could establish 675 trees on 10 acres of the quarter section. Unfortunately, the basic assumption was scientifically false, and land entries under this law were often fraudulent.

The sixth method for obtaining public land was under the provisions of the Timber and Stone Act of June 8, 1878. One could obtain title to 160 acres of land unfit for cultivation but valuable for the production of stone or timber by paying $2.50 per acre. This program was originally limited to California, Oregon, Nevada, and Washington, and was little used in Nevada.

The seventh method of acquiring public land, through the Desert Land Act, permitted entry provided the land could be irrigated. The act was never popular in Nevada. By the time it was passed, all the available irrigation waters had already been appropriated.

An eighth method of obtaining land was through the mining laws. Secondary ownership after mining was finished could be of great strategic value if the land contained stock water. An individual could locate a lode claim 600 1,500 feet at $5.00 per acre, or a placer claim not exceeding 160 acres. These claims did not necessarily lead to a land patent. A patent could be applied for if the mineral claim was productive. The Elko newspaper often published notices of applications by John Sparks or Jasper Harrell for patents on various mining claims.

The ninth method of obtaining public land was under the Coal Land Law and related to the development of the western railroad network. The railroads were eager to develop local sources of energy in the West to avoid shipping costs. This method of land entry was important in Wyoming, which had abundant coal deposits. More than 100,000 acres were patented in Wyoming under this act. This method was not so important in the Intermountain area because of the lack of coal deposits along the railroad, and only 1,600 acres of land were patented under this law in Nevada.

Land entry could also be obtained by using military-bounty land warrants awarded by Congress to soldiers for service in past wars. The warrants could be used to obtain land or could be sold to land speculators for cash.

Which method of land entry was most important in building the livestock ranches of the Intermountain area, and specifically in the Great Basin? The surprising answer is, none. The primary legal methods of obtaining patents to public lands were not satisfactory for building ranches because of their acreage restrictions. Every contemporary review of federal policies made from 1870 to 1900 confirmed this fact, but to no avail.[4] No changes were made in government land policies to accommodate ranchers' needs.

Probably the most famous of the late-nineteenth-century public land reviews were those made by Major John Wesley Powell. Among his recommendations were the following:

> The grasses of the pasturage lands are scant, and the lands are of value only in large quantities. The farm unit should not be less than 2,560 acres, the pasturage lands need small tracts of irrigable lands, hence the small streams of the general drainage system and the lone springs and streams should be preserved for such pasturage farms, the pasturage lands will not usually be fenced, and herds must roam in common. As pasturage lands should have waterfronts and irrigable tracts, and as residences should be grouped, as the lands cannot be economically fenced and must be kept in common, local communal regulations or cooperation is necessary.[5]

Powell felt that existing land laws were inadequate for settlement of irrigable lands.

Powell's report captured the inherent nature of the sagebrush/grasslands environment: irrigable lands and rangelands had to be tied together. Settlers needed small blocks of irrigated land to raise forage for wintering stock as well as extensive blocks of rangeland. The 2,560-acre blocks of range proposed by Powell were still much too small for the more arid areas, but the proposal was a step in the right direction. One of Powell's more radical proposals was abandonment of the rectangular system of land survey to allow land claims to fit the soils and topography of specific situations. If irregular shapes were allowed, the 80-acre irrigable tracts could fit available alluvial soils along streams.

Thousand Springs Valley, which constituted a major portion of the

Sparks-Tinnin ranches, provides a classic example of the stringing together of 40-, 80-, and 160-acre pieces of land astride alluvial and
irrigable soils. Thousand Springs Creek starts and ends in the checkerboard of the Central Pacific Railroad grant. As the creek swings north in
a great arc around Tony Mountain, it passes out of the checkerboard. In
this area the private land occurs as a narrow band along the stream. The
edges of the band are stair steps caused by the joining of surveyed rec-
tangles. Had Powell's suggestion been followed, a smooth band along the
streams could have been obtained.

Ranches in Nevada were built not by direct entry on the public domain, but indirectly through the purchase of state lands. More than one-
half of the deeded land in Nevada was originally obtained through purchase of state school lands.[6] When Nevada became a state in 1864, it
received a number of land grants from the federal government, including 3.9 million acres to be sold for school support. After 1848, every state
entering the union received two sections (16 and 36) in each township
to be sold for school support. An internal improvement grant of 500,000
acres was the second largest grant, followed by 90,000 acres for an agriculture college, 46,080 acres for public buildings, and 9,228 acres as
an indemnity grant.[7]

The initial demands for state land were not met from the 700,000
unspecified acres ceded by the federal government in the statehood
settlement. By 1871, timber, ranching, and farming withdrawals had
depleted most of the unspecified acres, leaving the specific sections 16
and 36 of state school land grants in each township. In 1873 the state legislature of Nevada asked Congress to exchange this grant for 1 million
unspecified acres, pointing out that sections 16 and 36 often occurred in
the middle of barren playas or on the top of rugged mountains. In 1879
the Nevada legislature again approached Congress, this time asking for
1.5 million acres in exchange for the original grant of sections 16 and 36.
This exchange was finally implemented on June 16, 1880, when a generous Congress authorized the transfer of 2 million acres of unspecified

land to Nevada and accepted the unsold acres of the original public school grant in return. Of the original 3.9 million acres in sections 16 and 36, slightly more than 63,000 acres had been sold.[8]

The government's 2-million-acre grant, sold to applicants in maximum units of 640 acres, was depleted in less than twenty years. The bulk of these state school lands were selected in Elko, Humboldt, Lincoln, and Washoe Counties, where there were large ranching companies. John Sparks became a major purchaser.

Between 1883 and 1893, 367,926 acres of state lands were sold in Elko County. Between 1893 and 1903, 230,808 acres were sold, far above total acreage for any other county in Nevada. Most state school lands were sold for a down payment of twenty-five cents per acre, with the entrant either meeting the credit provisions of the purchase contract or forfeiting the acreage. Acreage restrictions, land prices, and interest rates were eased by successive legislatures to make the land acquisition process less cumbersome and expensive. Ranchers and speculators often held the land for decades without making a single payment, and the state surveyor general and legislature silently permitted the practice. By 1902 only defaulted lands, often overgrazed and stripped of usable timber, were available to prospective buyers of state school lands.

From congressional land grants in 1862 and 1864, the Central Pacific Railroad received 5 million acres of public lands in Nevada. The railroad lands formed a checkerboard on the right-of-way, which paralleled the Truckee and Humboldt Rivers over much of their length and included a high proportion of the finest agricultural lands in the state.[9] The grant of public lands to the Central Pacific was more than double the state's grant for support of its schools. Nevada newspapers and public documents expressed little concern over the magnitude of the railroad grants until the middle 1860s.[10]

Settlers within Central Pacific's grant had the option of filing preemption or homestead claims under federal law and/or filing for 320 acres under the State Land Act. The registrar of the U.S. Land Office in Carson

City visited settlements along the Humboldt to accept entries locally.[11]

The Nevada State Board of Regents was desperately trying to support an embryonic school system during the 1870s. The only way to raise money was by selling grant lands. The Board of Regents set a minimum figure of $1.25 per acre except for the vacant federal lands within the railroad checkerboard, which were priced at $2.50 per acre. The state attorney general found that land application could be accepted without competitive bidding, and opened 700,000 acres of unspecified state lands to entry without advertisement of competition. By 1874 most of the 700,000-plus acres of unspecified lands had been sold.[12]

The 1880 congressional grant of 2 million unspecified acres enabled cattlemen to expand their landholdings at a time when both demand for beef and prices were rising.[13] This was significant for Nevada because the Comstock Lode's output declined after 1876, with a concomitant effect on the state's economy. In 1881 the Nevada legislature authorized applications for 320 acres from the state grants of 1864 and up to 640 acres from the 1880 grants. The 1885 legislature rewrote the basic land statute, reducing the interest rates from 10 to 16 percent and extending the repayment time from nine to twenty-five years. This was later extended to fifty years with 6 percent interest on the balance. In other words, ranchers obtained use of the land for 6.5 cents per acre per year![14]

According to historian John Townley, sales provisions enacted by the 1881 and 1885 legislatures were utilized to concentrate land among ranchers, with the State Land Office as an active participant. Assistance given to individual applicants by Land Office staff violated both the letter and intent of the land statutes but accurately reflected the attitude of the office during the twenty years required to disperse the two-million-acre school land grant. Applicants exceeding their legal limitation were openly advised to "have some member of your family who has not exhausted his or her right, or some person who will deed it to you, sign and return it to this office."[15]

Many of the largest land purchasers in Nevada, such as John Sparks,

John G. Taylor, Miller and Lux, W. N. M. McGill, and the Dangbergs, deposited funds in Carson City to be drawn upon to meet the numerous annual credit payments for state school grant lands. These firms had dozens of contracts on various parcels. The State Land Office personnel kept track of the required payments for large ranchers in return for a private fee.[16]

The private lands that John Sparks and Jasper Harrell acquired generally were restricted to areas that could be irrigated or to strategic springs. These areas were often rather narrow, like the land along the middle portion of Thousand Springs Valley. Near San Jacinto, where Trout and Shoshone Creeks join Salmon Falls to form a broad valley, Sparks and Harrell acquired all the available bottomland that could be irrigated.

The major exception to purchasing only irrigable land was John Sparks's acquisition of Gollaher Mountain. Located east of San Jacinto, Gollaher Mountain's eight-thousand-foot highland forms the divide between streams draining to Salmon Falls Creek and Rock Creek to the south and Goose Creek to the east. In a series of acquisitions from 1890 to 1896 that provide a study of nineteenth-century land policy, John Sparks acquired title to nearly a full township of high-elevation rangeland on Gollaher Mountain. This is one of the rare examples of a nineteenth-century rancher obtaining title to his summer rangelands. It is also surprising that he initiated these extensive purchases in 1890, a year when Nevada cattlemen were extremely short of cash.

After 1880 the Central Pacific Railroad offered land at 20 percent down and 7 percent interest over five years. Prices ranged from two to ten dollars per acre. Central Pacific held the land without patent from the federal government and untaxed by local governments. In 1885 Central Pacific's holdings were divided into natural ranges based on topography, with each range offered for lease as a whole, usually at 2.5 cents per acre. Some ranges were as small as 10,000 acres and some as large as 250,000 acres. The largest of all was the 500,000-acre lease on Thousand Springs Creek that went to Sparks and Tinnin.[17]

On March 11, 1889, the Nevada legislature required all preemptive rights to be recorded by December 12, 1889. A second statute made owners of livestock trespassing on private property liable to double the damages.[18] The first bill affected small landowners who had squatted on small water sources for grazing since the territorial period. Most had protected their improvements by preemptive claim but had never applied to purchase the land from either the state or the federal government. Now they were forced to make entry or pay for the land in full. Many small ranchers simply did not have the capital necessary to purchase the land on which they were squatting. Their only recourse was to sell to the larger ranchers. The second law allowed ranchers who owned watering points to refuse to allow livestock belonging to others to use the water.[19] The trespass law strengthened the control of the ranchers with deeded land over migratory sheep, but the fact that sheep could use snow for water on winter range negated the law.

From 1887 to 1899 the land statutes remained substantially unaltered. Terms were liberal, repayment provisions rarely enforced. Annual land entries declined from a peak of 400,000 acres to an average of 60,000 acres.[20] This period was marked by a national depression in the early 1890s, plus a prolonged depression in Nevada.

In 1899 the legislature finally asserted itself by requiring prompt payment on contracts. The State Land Office could receive overdue payments one year from the due date without the entrant's losing the land. After that, contracts in arrears became null and void. This resulted in voluntary forfeiture of hundreds of thousands of acres. By 1903 nearly a million acres had reverted.[21] In March 1900 the State Land Office advised each county that the school land grant was closed. The preceding twenty years had seen a period of economic depression for Nevada, but the two million acres of state school grant land had passed into private hands to found a vigorous cattle industry.

Ranchers often acquired land by using an intermediary for the actual entry. Cowboys were hired to enter on land and then turn it over to the employer for a small sum. Or the cowboy homesteaded the public land, then applied to have the homestead converted to a preemption claim. After conversion, the rancher paid the $1.25 cost of the land and a fee to the cowboy, who transferred title to the rancher.[22]

The Homestead Law made no provision for the government to take back land on which the settler had failed or become disillusioned after receiving patent. Ranchers often obtained land from banks and stores that had loaned money to homesteaders and then received the homestead when the settler decided or was forced to leave.[23]

Nevada represents the ultimate in federal ownership of lands among the adjacent forty-eight states. From the original cession until 1934, when President Franklin D. Roosevelt closed the remaining vacant lands to entry, only 6 percent of the available public domain, excluding railroad grants, passed into private ownership. There was often no legal way to obtain title to the acres of rangeland necessary to sustain livestock. Even if there had been, the ranch operations probably could not have survived the tax burden that ownership of such lands would have imposed.[24] The only option was to use the public lands and try to protect one's possessory grazing rights. John Sparks used the years from 1880 to 1900 to give his empire legal substance in terms of landownership. Although he owned only a fraction of the total rangeland he used, he gained ownership of irrigable land where hay could be produced and parlayed the package into an empire.

CHAPTER 6

John Sparks
Capital, Credit, and Courage

John Sparks had a good thing going in Wyoming. Unfortunately, there was a limit to the rangeland available east of the Rocky Mountains, and by the end of the 1870s the ranges were almost fully stocked. Sparks did not come empty-handed to the Intermountain area to exploit the virgin range. He came with ready capital, credit, and expertise. He brought an impressive appearance, poise, the ego of a cattle king, and an abundance of nerve.

Sparks later told H. H. Bancroft:

In 1881 I went to Nevada and formed a partnership with the John Tinnins of Elko County. The range is known as the 1,000 Springs Valley Ranch and Range. In 1883 our firm purchased the Barley Harrell property, on the Salmon and Snake rivers in Nevada, and the territory of Idaho. The entire property now owned by this firm is known as the Rancho Grande. The great bulk of the stock is blood stock, being an admixture of Hereford and Shorthorn. The firm is a member of the National Stockgrower's Association. We carry on our business without a ledger of any kind; don't even have a bookkeeper. In fact, I keep my accounts in my head. We ought to and probably soon will have a bookkeeper. We now have between 80,000 and 90,000 head of cattle and are almost land poor, as the expression goes.[1]

In his new ventures of the 1880s Sparks joined with John Tinnin. Colonel John Tinnin's name is frequently mentioned in connection with significant events in Nevada, but concrete facts about his background are difficult to find. One source indicates that he, like Sparks, was the son of a Mississippi planter, and that he fought Indians as a Texas Ranger during the Civil War.[2] In the late 1860s Tinnin was employed as a livestock commission agent handling the sale and delivery of Texas cattle to the Intermountain area for the Ingram Company of Salt Lake City, Utah. Tinnin owned a classic steamboat Gothic house at 1220 Austin Avenue in Georgetown, Texas, which he purchased at about the same time that John Sparks became well established in that growing town. Tinnin's home was noted for its beautiful furniture and its parrot. The vocabulary of the parrot frequently shocked visitors.[3] Mrs. Harold G. Scoggins of Georgetown remembers, "It was an interesting old home. When I was growing up here, it was furnished in rosewood furniture, even the piano. It had a circular stairway from the kitchen to the master bedroom. There was a square tower at the front—over the entrance. Next to my dad I loved Mr. Tinnin, in spite of the fact I was punished for using his 'cuss' words."[4]

Sparks and Tinnin purchased their first ranch in 1881, "Old Bill" Downing's H-D Ranch on Thousand Springs Creek. Old Bill had left an immigrant train to found the H-D. Though not a large ranch, it was in a strategic location for control of the upper reaches of Thousand Springs Valley. On November 6, 1881, the *Elko Independent* announced that the Sparks-Tinnin partnership had purchased the Jasper Harrell and Armstrong property at Tecoma, where Thousand Springs Creek flows north of the Pilot Range onto the Bonneville Salt Flats. The property was valued at $150,000, but the purchase price was not announced. The sale did not affect Harrell's holdings on Salmon Falls and Goose Creek, or his extensive holdings in Idaho.[5]

On June 15, 1883, the *Elko Independent* reported the sale of the remaining Jasper Harrell properties to Sparks-Tinnin for $900,000: $100,000 down and the balance due in eight yearly installments of $100,000 each, with a 4 percent interest on the balance. The sale included thirty thousand head of cattle, a large number of horses, and extensive rangeland said to be one hundred miles square. Harrell customarily branded ten thousand calves on this range and the previous year had shipped $120,000 worth of beef.[6]

Why did Jasper Harrell decide to sell his holdings, and on terms so favorable to the purchasers? Was he tired of overseeing the widespread operation? Certainly, he did not need the money. He had accumulated enough wealth to last his lifetime. The wily old forty-niner may well have realized that the ranges were overstocked and the beef bonanza was vulnerable. He had sent his son, Andrew Jasper (A. J.) Harrell, to Nevada in the late 1870s to learn the cattle business. A. J. was a graduate of Heald Business College in San Francisco. Did Jasper sell so he and his son could concentrate on banking and real estate interests in California?[7] The reasons for the sale were probably many.

In November 1883 the *Elko Independent* printed an article with a headline calling Sparks and Tinnin "the cattle kings of the west." The seventy thousand head of cattle they owned and seventeen thousand calves they branded annually made them the "largest ranchers in the west." Swan Brothers of Wyoming ran more cattle, but that was a stock company rather than individual ownership. Judge Carey of Cheyenne, who ran thirty thousand head under his individual ownership, was reported to be the second largest ranch operation in the West. When the *Weekly Drover's Journal* of Chicago reprinted the article, Sparks and Tinnin got national recognition.[8]

In the 1880s money for ranch operations was hard to obtain and expensive to borrow. John Clay once had trouble getting a $5,000 operating loan when his ranch was valued at $500,000, debt-free, with five thousand steers to market that fall.[9] Interest rates were quoted at 1 percent per month and 15 percent per year. One eastern banker who loaned money on Nevada cattle came west to verify the collateral. He rode the range for days, covering several hundred miles in a buggy, but saw only a handful of cows representing collateral. He commented that he "might as well loan money on a school of fish in the Pacific Ocean."[10] During good years, most cattlemen liked to count their money in cattle. When dry years and low prices occurred, they found it difficult to convert cattle into cash sufficient to cover expenses without impairing operations.[11]

Sparks-Tinnin had $100,000 plus interest to pay annually on a ranch operation that had grossed $120,000 in 1882, the highest cattle price year of the decade.[12] It is possible that they had inside knowledge that Jasper Harrell actually had more cattle than he realized. C. W. Hodgson considered this a classic "book sale."[13] Ranches were sold based on the number of head carried in their books. These numbers were often greatly inflated, leading to bizarre incidents such as the Scottish accountant who painted identifying marks on cows in an attempt to balance the cows on the range with the number of cows in the books. In the case of Sparks-Tinnin it could have been a reverse "book sale," with the number of cattle purchased lower than the actual number present. John Sparks was once asked how many cattle he owned. "We leave those matters to the assessor and he comes around once a year," he answered. "It is an unwritten law that a cattleman never talks of the size of his herd."[14] During the course of his operations in Nevada, Sparks claimed a variable number of cattle. He may have adjusted the number to suit the occasion—most ranchers did—but in truth, he probably did not know the actual number.

The favorable political climate was a major factor aiding the growth of ranching in Nevada during the nineteenth century. One of the better-known cowboy governors was Lewis Rice Bradley, the second governor of Nevada, who served from 1871 to 1878.[15] Bradley drove cattle from Missouri to Stockton, California, in 1852.[16] He led what was known as the "Bull Block" in the state senate, so named because the group represented the interests of ranchers at a time when Nevada politics was focused on the Comstock miners. Bradley ranched in California until 1862, when an exceptionally dry winter led him to take his operation to the Nevada sagebrush range. He established himself first in central Nevada to furnish beef to the Austin mining district. He later moved from central Nevada to Pine Valley and the South Fork of the Humboldt River in Elko County. The governor's son, J. R. Bradley, in partnership with the Russell family, developed a huge ranching operation extending into the Snake River valley and lying just west of Sparks-Tinnin holdings. They had common fall roundups, or rodeos, where the present-day Twin Falls, Idaho, is located.[17] Nevada cowboy governors after Bradley included Jewett W. Adams (1881–86) and Reinhold Sadler (1896–1902), who also had large ranching interests.

The primary market for Nevada cattle during the 1880s was the growing population of California. Relatively few meatpackers controlled most of the market. In the San Francisco area, Miller and Lux began to dominate the markets and have a major voice in establishing live-beef prices. Henry Miller applied to the California legislature for a butcher's reservation in south San Francisco and a guarantee that butchering might be carried on there for ninety-nine years. Then he built a slaughterhouse wharf big enough to accommodate every butcher in San Francisco. Although the wharf was destroyed in the 1906 earthquake, the supporting piles remained for years because of the thousands of tons of offal dropped under the wharf.[18]

Self-sufficient in beef production by 1870, Nevada supplied an estimated thirty thousand animals annually to the California wholesale butchers.[19] Later in the 1870s, a group of Reno and Winnemucca businessmen decided to butcher their own cattle and ship dressed beef to California. Thirty carcasses, rather than eighteen to twenty live animals, could be shipped per railroad car. The California Wholesale Butchers Association broke the new Nevada company by refusing to sell to any retailer in California who bought the Nevada-dressed beef.[20] By 1880 Nevada was supplying one-half of San Francisco's beef. In 1884 it was estimated that the San Francisco market required 250 head per day. Long trains of cattle cars filed through Reno loaded with Elko County cattle on their way to the Bay Area market.[21]

Freight rates from Halleck, near Elko, to Chicago were $260 per car; the rate from Halleck to San Francisco was $120. The trip east took at least eleven days, with feed and water necessary along the way.[22] With this differential in mind, the California Butchers Association quoted lower prices than eastern markets. When the price of live cattle dropped after 1885, the San Francisco market was glutted and Nevada ranchers turned to eastern markets to find an outlet for their product.

John Clay became acquainted with Sparks and Tinnin in 1885 when he bought three thousand steers from them delivered at Rawlins and Rock Creek, Wyoming. The two-year-old steers cost twenty-seven dollars; the three-year-olds thirty-five dollars.[23] The same year Sparks-Tinnin shipped nineteen hundred head of young mixed stock to J. G. Pratt, the land agent for the Union Pacific Railroad, delivered at Antelope on the Cheyenne River.[24] In October 1887 the Elko paper reported that Sparks and Tinnin had just returned from marketing cattle in Omaha.[25]

Shipping by railcars was the innovation that made the beef bonanza possible. It was especially important in the Intermountain area because the region was rimmed with mountains, forests, and deserts that made it virtually impossible to walk fat animals to market. Despite their value, railroad stock cars were hard on cattle. Sudden starts and stops caused injury. Cattle that went down were usually trampled by the other stock. Losses of cattle shipped from Colorado or Wyoming to Chicago were estimated at one and a half per thousand shipped.[26] In 1872 the American Humane Education Society offered a one-hundred-dollar reward for the best essay on the transportation of animals. George T. Angell, writing under the pen name Litera, won the prize, and twenty thousand copies were printed and distributed. In 1873, after prodding by the American Humane Education Society, Congress passed the Twenty-eight-Hour Law over the bitter objections of stockmen and meatpackers.[27] This law required all stock to be removed from trains at least every twenty-eight hours for a period of five consecutive hours for feed, rest, and water. During the 1860s it was customary to crowd thirty steers into a railcar. Federal regulations passed to promote more humane treatment required room for one-third of the animals to lie down. Acceptable carloads were based on live weight of the animals being shipped: 700-pound animals were twenty-three per car, 1,000-pound animals were twenty per car, and 1,400-pound beefs were sixteen per car. Ranchers protested this law as well, claiming that it provided room for cattle to go down and be trampled, while the densely packed cars actually protected the animals.

Loading cattle for shipping was a tense business. It was hard work done under supercharged conditions against specific deadlines. The railroad would hold the cars on a siding for only a short time without an additional surcharge. The rancher had to request the cars and plead with the agent to have them spotted on the siding at the correct time. The steers had to arrive at the shipping corral in sequence with the cars. Wild range steers, half-broken saddle horses, locomotives belching clouds of steam, and shrill whistles did not mix well. A toot on the locomotive whistle at the appropriate time by a sly engineer would set off a rodeo that might end with several cowboys on the ground.

Shipping corrals were located in Elko, Carlin, Tecoma, Wells, Deeth, Halleck, Palisade, Red Rock, and Iron Point along the Central Pacific route through Elko County. The steers were held in bunches on open ground until it was time to load them. Then, sufficient animals to fill the corrals were cut out and forced through the gate. Inside, a crush corral funneled steers into the loading chute. The corrals were either choking with dust perfumed with cow droppings or ankle-deep in sucking mud. The steers had to be forced up the chute with shouts and curses. If words did not work, cowboys poked poles through cracks in the chute and jabbed them in the ribs.

The classic "wide horns and narrow chute" cartoon shows cowboys trying to force a steer with eight-foot horns through a four-foot-wide railcar door.[28] If a dog barked at the wrong time, the steers would balk or try to turn in the chute. The range-raised steers would turn and fight if provoked. Men on foot in the corrals were in constant danger.

After the last car was loaded and rolling down the tracks, the cowboys found themselves in town with the opportunity for a spree. In Elko, they headed for Dobe Row, a block and a half of adobe houses that formed a noted red-light and gambling district. The rancher could now relax, too, knowing that cash money was on the way from the sale of the shipped animals.

In 1880 the Tenth United States Census interviewed nineteen Nevada ranchers to determine the composition of range cattle herds. The census was taken one year before Sparks-Tinnin came to Nevada, but it provides insight into range operations of that decade. The nineteen herds surveyed contained 94,786 head, broken down as follows:[29]

	Number	Percentage of Herd
Bulls	1,422	1.5
Cows	28,439	30.0
Three-year-old steers and beefs	11,861	12.5
Two-year-old steers	14,219	15.0
Yearlings	19,907	21.0
Calves	18,948	20.0

It was customary during the 1880s to allow bulls to range with cows throughout the year. This practice led to the untimely dropping of many calves; severe late-winter storms often proved fatal to both cow and offspring. The percentage that survived to yearling age was estimated at 66–80 percent in 1880 for Nevada. The estimated average annual loss among cattle more than two months old was 6 percent, arising from disease, winter and spring storms, snakebites, wild animals, theft, and poisonous weeds.[30] It is interesting that the ranchers who responded to the 1880 census failed to list starvation as a cause of death.

Theft of range cattle was always a problem. Colonel E. P. Hardesty of

Elko County gave instructions to his cowboys regarding anyone they caught killing his cattle: "If he stole it to eat, tell him to enjoy it and bring me the hide. If he stole it to sell, bring me his hide."[31] Sam McIntyre, who ranched off the North Fork of the Humboldt, ran Galloways, Scottish Highland cattle. Because of the uniqueness of the breed, he did not bother to brand his stock until theft became a serious problem. A reporter once asked Sam who else in Elko County was running Galloways, and Sam replied, "Everyone who has a horse."[32]

One Elko County ranch foreman was much less charitable. When he caught a homesteader skinning a company cow, he forced the man to nail the hide to his shack with the brand showing so that everyone who passed by would know he was a thief.[33]

The 1880 census estimated the "Cost of going into the ranching business on the Humboldt River in Nevada" as follows:[34]

2,000 three-year-old cows	@ $12.50	$25,000
100 two-year-old bulls	@ 50.00	5,000
25 saddle horses	@ 50.00	2,250
2 work horses	@ 100.00	200
1 wagon and harness	—	120
Ranch building, saddles, etc.	—	2,000
TOTAL	—	$34,570

The annual expenses of a Humboldt River Ranch were listed as follows:

Five cowboys	$40/month for 8 months	$1,600
One cook	$30/month for 12 months	360
Two cowboys	$40/month for 2 months 1	60
Provisions for men	$12/month	720
Taxes on cattle	—	450
Taxes on horses	—	30
Taxes on ranch improvements	—	27
TOTAL		$3,347

Experienced herdsmen estimated their annual profit on capital invested at 20–30 percent if death losses were not greater than 5 percent.[35]

In 1885 the average cost of raising a steer, including interest on capital invested, was estimated by the larger stockowners at $0.75–1.25 per year. Thus, a large four-year-old steer ready for market cost $4–5 to raise.[36] During most of the 1880s, prices stayed in the $15–20 range, leaving a considerable profit margin. That changed at the end of the decade. When 1884 brought the second depression since the Civil War, the New York firm of Grant and Ward failed, with repercussions for western stockmen. And later, in 1886 and 1887, severe winters in Wyoming created panic in the livestock market.[37]

Only one-third of the range cows produced a calf each year. Often, calves were left on the cows until they weaned themselves. The big sucker calves prevented the cows from coming into heat or conceiving. Some ranchers tried to improve the quality of their stock during the 1880s, but that was difficult to do when several operators were ranging cattle on common range. The stockman who turned out superior bulls on the common ranges shared the benefits with all the brands.[38]

Sparks-Tinnin apparently made some attempts to improve herd quality during the 1880s by castrating bull calves, introducing improved bulls, and reducing stock numbers on the range.[39] The reduction in stock may well have been aimed toward making their mortgage payment, however, rather than an effort to improve the range.

Many observers recognized that the ranges were being overgrazed. On December 4, 1886, the editor of the *Carson City Morning Appeal* called for "appropriation of state funds for research to find ways to seed and restore the range."[40] But half a century would pass before large-scale restoration techniques were put into practice. In 1885 a special agent of the Bureau of Animal Husbandry reported, "Cattlemen are warning that the western ranges are overstocked and petitioning Congress to lease . . . public domain [lands] at 1¢ per acre."[41] Not everyone recognized the severity of the problem. The newness as well as the immensity of the ranching enterprise left American ranchers without standards by which to gauge either the security of the roaming herds or the capacity of the forage to hold up under continued intense utilization.[42]

The Spanish contributed the techniques required to work cattle in the open without fences and corrals. Along with that knowledge came the vocabulary of ranching in the western United States plus the lariat, a type of saddle, chaps, and the sombrero.

In 1880 the foremen of many of the large ranches in Nevada, eastern Oregon, and southern Idaho were called *mayordomo* rather than foreman. The cowboys were *vaqueros,* and there were about three for every thousand cows. The agrarian culture in California before the Mexican War was akin to that of fifteenth-century Castile or Extremadura. The invasion by Anglo settlers, the breaking up of huge ranches, and the rise of dryland cereal production spelled the end of that way of life in Old California and sent the Spanish cattle to the sagebrush/grasslands.

Cowboys' use of rawhide for their tools was an important part of the Spanish heritage. The rawhide lariat, whose name derives from the Spanish *la reata,* became a symbol of this technology.[43] The pliable hide of a relatively young animal was preferred. A fresh hide was allowed to dry for four or five days. If a dried hide was used, it was soaked in water with wood ash added to loosen the hair and soften the hide. The hair was stripped from the hide with a bit of broken glass. Each lariat was braided from four, or occasionally eight, strands. The cutting of the strands began in the center of the hide and worked outward. A sharp, thin-bladed knife was used, plus a wooden gauge to control thickness, usually three-eights of an inch. Strands for a seventy-foot rope had to be ninety feet long to allow for plaiting. The braiding took many hours. The four strands were tied together, and the weaving began over one, under one, right and left, drawn tight with no slack. At the end, the strands were worked back to form a loop, or hondo.

After braiding, the lariat was stretched with a heavy weight tied to one end and rubbed with tallow. It was twisted and stretched until it was perfectly round. The lariat had to be kept dry, and it was not as strong as manila grass or sisal rope. Rawhide rope was not strong enough to stand hard and fast tying, so a roper could not tie the rope to the saddle horn before roping an animal. The roper had to take a dally, or twist the rope around the horn, and let out the slack. Ropers were sometimes identified by their lack of thumbs. A thumb stuck in the dally when the cow hit the end of the rope was removed with surgical precision.

For the professional cowboy, a snapped rawhide lariat was no laughing matter. Louis Harrell, Jasper's nephew, was roping wild horses in the O'Neil Basin in 1896 when a rawhide rope broke and hit him in the eye, causing him to lose the sight in the eye. He was roping the horses to roach their manes and tails and sell the hair to prisoners at the state prison.[44]

Hair, or "mecarty," ropes were sometimes used for hackamores. Hair rope making became a highly developed art in the Intermountain area. Reins and headstalls became a specialty, and white and black blends to provide a salt-and-pepper effect were common. This Spanish-heritage art form was practiced by Indian craftsmen as well as prisoners at the Nevada and Idaho state prisons.

The highest state of rawhide craftsmanship was the quirt, the short whip that hung from the saddle. It was made by braiding six to twenty-four strands of rawhide over a wooden or metal center. The top of the quirt was finished in an elaborate Turk's-head design. Confederate cavalrymen rode into battle with weighted quirts as auxiliary weapons.

Each section of the Intermountain area had individuals or families known for their work with rawhide. In the Bruneau Valley of Idaho, for instance, Joe Samora, who came with the Longhorns in the 1860s, became well known for eight-strand reins and sixteen-strand works of art.[45] On the North Fork of the Humboldt, the vaqueros of Pedro Altube's giant Spanish ranch included several experts in the rawhide art form.

A ranch headquarters in the 1880s was just a point of departure; livestock operations were mobile functions done on the open range. There were two roundups each year, in spring and fall. The spring roundup was primarily concerned with branding, marking bull calves, and earmarking slick-ear calves. The fall roundup picked up slick-ears missed by the spring passage and sorted out the fat beef that would be sent to market. The roundup operation evolved from the trail outfits that drove Longhorns from Texas across the plains to the new ranges in the Northwest.

Roundups were centered on wagons. Usually there were about twenty cowboys to a wagon. Each man had five to ten horses, some broken, some green; he was expected to have the green horses broken by the end of the roundup. Some outfits hired a special bronco buster who worked on the green horses. The wagon, driven by the cook, formed a mobile headquarters for each roundup. It would move to a given place where water and horse feed were available. The cowboys would gather cattle and work them in a given holding area nearby, usually a large, level flat. They ate and slept at the wagon until all the cattle were worked in that area, then the entire operation moved on to the next location. The cowboys slept in bedrolls made of several blankets or quilts rolled in a canvas tarp and tied with rope. They were bulky—twice the length and two or three times the diameter of today's rolled sleeping bags.

The horse wrangler was responsible for the extra horses and had to be up well before the regular crew to bring the horses into camp. Usually a mare was fitted with a bell to help the wrangler locate horses in the dark. A wrangler's axiom was "hear the bell and sleep well."

Sparks-Tinnin usually operated two roundup wagons. One worked the Thousand Springs Valley ranches in Nevada, and the other worked the Snake River side of their range. Representatives (reps) of neighboring large ranges also went along and were responsible for identifying and branding calves that belonged to cows branded with their company's brand. Likewise, Sparks-Tinnin reps traveled with the Bradley and Russell wagons and other large roundups.

Once the cattle had been segregated by ownership, branding could begin. Calves were roped and dragged to the branding fire. On some roundups it was customary to rope by the hind feet rather than the neck. Calves caught in this manner were easier to throw and hold for branding, but a higher level of roping skill was required. A cowhand on the ground grabbed the roped calf by the flank and jerked it off its feet. While the horseback roper kept the rope taut, the cowboy grabbed the hind leg to keep the calf on the ground. The hot branding iron was applied; the sickening smell of burned hair filled the air. The appropriate earmark was made. If the calf was a male, it was castrated. Often the testicles were saved on a hot shovel blade at the branding fire and hungry hands could pick up a hot Rocky Mountain oyster. Huge numbers of cattle were worked at some of the roundups in Elko County. The Juniper Basin rodeo averaged ten thousand. The workday began at 4 A.M., and sixteen-hour days were frequent.[46]

A mess wagon cost approximately $125. The majority were "three to three and one-half inch" wagons, a reference to the diameter of the axle where it entered the hub. The Bain, manufactured at Kenosha, Wisconsin, and the Peter Schulter from Chicago were favorite makes of wagons.[47]

The chuck box, a sturdy cupboard built into the rear of the wagon, never received the dignity of a patent, but it became standard equipment on the range. The chuck box was two to three feet deep, and its perpendicular front was about four feet high. The rear wall of the box was hinged at the bottom so it could be swung down to form a worktable.

The inside was fitted with partitions, shelves, and drawers, and two doors folded snugly over the partitions to hold everything in place while the wagon was on the move. Each item had its place. The larger divisions were for the sourdough jar or keg, a partly used sack of flour, and bulky utensils. There were drawers for tin plates, cups, spoons, knives, and forks. Salt, pepper, soda, and baking powder were kept in tins with tight lids. Every cook reserved a drawer for purgatives and cure-alls like quinine, calomel pills, black draft, and horse liniment, the latter to be used on man or beast. Reserved for the cook's private use was a bottle of whiskey. Heavy supplies like flour, bacon, molasses, coffee beans, and canned goods were carried in the bed of the wagon. Attached beneath the chuck box was another, smaller box with a hinged door for Dutch ovens, pots, and skillets. One side of the wagon carried a water barrel, often wrapped in a wet gunnysack to keep its contents cool.

The cook's equipment also included a "fly," a canvas sheet that could be stretched above the end of the chuck wagon to make shade and shelter for the cook. Every cook needed a "gouch hook," or pothook, to lift the heavy lids off his cooking utensils. There was also a supply of fire hooks, iron rods from which cooking vessels were hung over the fire. The cook's most important piece of equipment was his Dutch oven, a very large, deep, thick iron skillet with three legs under the bottom and a heavy lid with upturned lip that could also be used as a skillet. Roundup cuisine was based on Dutch-oven cooking. The oven either sat on a rack placed over the fire or was buried in the coals.

If the crew was large and the weather cool, a steer would be killed for fresh meat. The first night's meal would be fried liver and onions, with Dutch-oven biscuits or pan bread. During warm weather, when fresh meat would spoil, the crew was stuck with home-cured ham, shoulder, or sowbelly.[48]

A sourdough keg kept in the wagon provided starter for the biscuits. Sourdough bread, which could be made without commercial yeast, was among the most famous of all western foods. The cowboy preferred sourdough bread to any other. The cook's most particular job in preparing for the start of a roundup was to secure the proper keg for his sourdough mixture. Most cooks would defend their sourdough kegs with their lives.

When he was ready to make bread, the cook poured flour into a large pan until it was two-thirds full, made a deep impression in the center, and poured in his sourdough batter. Next he added a teaspoon of soda dissolved in a little warm water and a small amount of salt and lard. As he stirred, the cook worked the dry flour from side to side, being careful to distribute the soda and shortening thoroughly. He then drenched his worktable with flour and kneaded the dough thoroughly. While he was preparing the dough the cook was also preparing the baking fire. The red-hot coals had to burn down to the proper temperature.

When the dough had been kneaded to the right consistency, the cook put a generous portion of lard in the Dutch oven to melt. The range cook had no use for a biscuit cutter. When the grease had melted in the Dutch oven, he merely pinched off pieces of dough somewhat smaller than an egg and rolled them into balls between his palms. He turned the biscuits in the grease, coating them on all sides to prevent their sticking together. As he placed the biscuits in the oven, he jammed each tightly against the others. The tighter they were packed, the higher they would rise and the lighter they would be.

The oven was then placed near the fire for thirty minutes to allow the biscuits to rise while the cook went about other preparations for the meal. When the other items were nearly ready, the cook placed the oven on the coals and covered it with more hot coals. The bread was better when fewer coals were used on the bottom and more on the top; the crust would be brown and the center tender.

Beans were a staple at all meals, including breakfast. No range cook would start his roundup wagon without a good supply of dried beans. Most cooks preferred the brown-spotted pinto beans. The black iron pot of beans cooking, with miniature geysers throwing up little jets of steam as the beans bubbled over a slow fire, was enough to excite any cowboy's appetite. Beans were usually cooked at least five hours over a slow fire. Pieces of dry salt pork were dropped into the pot for seasoning.

After the cook had the rest of the meal under way, he placed the coffee-pot, two-thirds full of cold water, on the coals to boil; a three-gallon to five-gallon pot was the standard size for ten to fifteen men. When the water boiled, he dumped in the correct quantity of ground coffee. After it had boiled to strength, he dashed in a little cold water to settle the grounds. The sight of a wide-bottomed, smoke-blackened coffeepot on the coals with the brown liquid bubbling down its sides was a picture to warm any cowhand's innards.

One favorite cowboy meal was a stew made from liver, brains, heart, tongue, marrow gut, sweetbreads, kidney, onions, and salt and pepper simmered together in a huge iron pot. For some reason this was often called "son of a bitch stew." After two or three days of reheating, all the internal organs tended to lose their individuality.

When the range cook prepared meat for a meal, he sliced off a sufficient number of steaks and tenderized them by pounding with a hammer on the back of a heavy knife laid over the meat. The cook cut suet into small pieces and put a handful into a hot Dutch oven to render. When the suet had cooked down, he fished out the cracklings that were left. The slabs of steak were salted, covered with flour, and dropped into the sizzling fat, and the lid put on. The steaks were always cooked well done. A special treat was to sit around the campfire at night and roast beef on a willow stick.

High on a cowboy's list of luxuries was freshly baked pie. The range cook baked his pies in the Dutch oven. The dough was rolled out with a beer bottle, placed in a greased pie pan, filled with stewed fruit, and covered with the top crust, which was trimmed with a knife and then scalloped around the edge with a fork to seal the top and bottom crusts together. On the top crust, to allow steam to escape, he usually cut the company's brand. Spotted dog—boiled rice with raisins added—was another favorite.

Cowboys bought their own clothes and equipment. Stetsons, costing from $7.00 to $30.00, were favorite hats. California spurs with twin bells on each shank were popular, too. They cost from $5.00 to $50.00, depending on the amount of silver used. "Noisy" shirts in colors like purple, red, and orange added color to the outfit. A canvas slicker covered with fish oil, cost $3.25, kept out the rain. A horse was not really "broke" until the cowboy could mount it while wearing his slicker. The average cowboy paid $24.00 for a Spanish bit, $5.00 for a bridle, $25.00 for a horse, $60.00 for a saddle, $10.00 for a saddle blanket, $2.50–5.00 for a quirt, $10.00 for spurs, and $10.00 for a hat. In all, his personal equipment cost about $150.00.[49]

Cowboy life at the end of the nineteenth century was a brief interlude in the nation's history, but it left a lasting imprint on the American identity. President Teddy Roosevelt reflected on his experience in the American West: "We who have felt the charm of the life have exulted in its abounding vigor and its bold, restless freedom, will not only regret its passing for our own sake, but must also feel real sorrow that those who come after us are not to see, as we have seen, what is perhaps the pleasantest . . . and most exciting phase of American existence."[50]

John Sparks and John Tinnin built a great and far-flung cattle empire on the sagebrush/grasslands. It reached from Wells to Pilot Peak on the south and to the Snake River on the north. They owed money on huge mortgages, and their range was overstocked and overgrazed. The market for beef was severely depressed after the hard winters of 1886 and 1887. Despite all these problems, John Sparks was positive and confident when he told H. H. Bancroft that the firm had no books, that all its accounts were kept in his head.[51]

CHAPTER 7

White Winter
White Hills of Bones

Rarely does a single climatological event alter the plant and animal ecology or change the social and economic structure of a wide geographical area. However, such a far-reaching and dynamic event was the devastating winter of 1889–90 in the sagebrush/grasslands of western North America.

Herdsmen are traditionally resistant to change, and they thrive on the repetitive cycle of new grass, calving, branding, marking, and marketing. Traditional systems of management often persist long past their time in defiance of economic laws. Although it was less than fifty years old, the practice of open-range management was just such an outdated system, and the winter of 1889–90 brought a white wave of disaster and hardship as a result.

In hindsight, it is easy to suggest that late-nineteenth-century ranchers had been offered ample evidence that open-ranging livestock without conserving forage for winter feeding was an invitation to disaster. At the time, however, the open range appeared to be an exciting, romantic, and economically attractive industry. Technically, Anglo-Texas pioneer ranchers were trying to exploit the grazing resources of a semiarid en-

vironment using techniques evolved by Spanish herdsmen in a Mediter-
ranean environment.

Frank Dobie devoted a volume to the anatomy, culture, and color of
the Longhorn. In a footnote he offered a classic ecological maxim: the
millions of Longhorns that were accumulated in Texas before and during the Civil War were not "free roaming animals of the plains," they were
free-roaming creatures of the oak woodlands, brushlands, and woods of
southern and eastern Texas. These areas were much more favorable winter environments than the open plains. When the Texas Longhorns were
driven to the open grasslands of the plains and prevented from drifting
south with the winter storms, they could not adapt and could not survive
without the aid of man.[1]

If the open-range system the Anglo-Texans borrowed from the Span-
iards had been moved gradually northward until excessive winter losses
were encountered and then stopped, the disaster might have been
avoided. That was not the case. Ranchers continued to push north and
west despite serious losses such as cattle suffering and dying in large
numbers as they wintered on the prairies around Kansas railheads. Year-
to-year variance in winter severity, regional differences for any given
winter, and the changing relationships between range condition and the
amount and quality of forage were among the reasons why ranchers were
slow to appreciate the risks involved in open wintering.

The winter of 1885–86 was severe on the southern Great Plains but
relatively mild in Wyoming and the new Northwest range states. Winter
losses in Kansas, Indian Territory, and the Texas plains were commonplace. In August 1885 more than 200,000 head were forced out of Indian
Territory by an edict of President Cleveland and moved to already heavily
grazed ranges of adjoining states. They were about to face one of the most
severe winters in history. George B. Louis, in written testimony for the
Nimmo Report, reported losses in some parts of Texas for 1885 and 1886
as approaching 30–40 percent.[2]

A factor contributing to the large winterkill on the southern Great

Plains in 1885–86 was the extensive fencing of the range with barbed wire. Patented in 1874, barbed wire was quickly accepted, and by 1880 annual production had reached forty thousand tons. The fences in northern Texas were designed to curb the southward movement of range cattle before winter storms. Longhorns differed from American bison in a basic instinct that had great influence on their winter survival. The bison faced or drifted into the storms, but the Longhorn turned tail and drifted south. In the winter of 1885–86, barbed-wire fences hindered the southward drift of the Longhorns, which piled up and died by the thousands.[3]

After the 1850s, when it was discovered that cattle could winter on the high plains of Wyoming, there was an escalation of winterkills. Overgrazing had already led to declining range conditions at this time.[4] John Clay clearly recognized that the higher losses resulted from wintering livestock on poor range, but this did not prevent him from joining the rush to disaster. Why did herdsmen whose heritage and training had taught them that animals should be maintained in fenced areas or under constant herd supervision suddenly turn thousands of animals loose in the open range? The Mormon settlers had not succumbed to this notion. Their first agricultural communities established in Utah in the 1850s followed the northern European practice of communal herding during the day and returning the animals to the village at night.[5]

Spanish ranchers had been very successful open-ranging cattle in balmy California. The technique was sold to the new wave of ranchers in the northwestern plains, which were not at all balmy. One of the most enthusiastic salesmen was Dr. Hiram Latham, a medical doctor for the Union Pacific Railroad stationed at Laramie, Wyoming. In the 1860s Dr. Latham wrote a series of letters to the *Omaha Daily Herald* describing the Laramie plains and their potential for livestock. These ideas were compiled into a pamphlet that was widely distributed by the Union Pacific Railroad to attract settlers. Dr. Latham described the Laramie plains as a year-round paradise for livestock: "The grasses are self-curing, and sheep and cattle live and thrive year round without other food or shelter than that afforded by nature." He cited areas of supposedly similar climate in South America and South Africa where cattle were open-ranged, reinforcing his arguments with quotations from the Bible. "The secret of these great herds of cattle, horses, and sheep for so many centuries," he continued, "is *winter grazing*. I speak of the grazing in all these countries to show that the idea of cattle grazing in winter in the latitude and altitude of these plains is not new, but as old as the history of man."[6]

The winter of 1886–87 again denied Dr. Latham's prophecy. Total rainfall for May, June, and July 1886 was 2.55 inches at Cheyenne, compared with an average of 5.15 inches for these three months during the twelve previous summers. The number of calves counted had been very low at the spring roundup in 1886 in Wyoming because late spring storms had decimated the calf crop. Montana ranchers attributed some of their losses to cattle grazing poisonous plants. Granville Stuart blamed losses on the drought and overgrazing of the desirable forage species.[7]

Wildfires were commonplace in Wyoming during the summer of 1886. The fires consumed grass vital for the coming winter. Big fires occurred along the foot of the Judith Range and on the Musselshell in Montana, and the sky was often obscured by smoke and dust. John Clay considered the summer of 1886 the driest he had experienced in thirty-five years of livestock activity in Wyoming. He rode over the south-central Wyoming range and saw "scarce a blade of grass." The same conditions prevailed, he said, on the Belle Fouche, Little Missouri, and Powder Rivers. Streams as large as the Rosebud ceased to flow. Because of drought and overstocking, range animals approached the winter in poor condition. Many were recent arrivals from Texas new to the ranges on which they were to be wintered.

Between January 28 and January 30, 1887, a record blizzard swept down from the north. Eastern Montana ranchers thought the Arctic had suddenly enveloped them. It was -46°F. Cattle drifted before the storm but found no forage to sustain them. Even fat steers froze to death along the trails. Inhabitants of Great Falls, Montana, looked out through the swirling snow one morning to see the gaunt and reeling leaders of a herd of five thousand cattle that had drifted south from the frozen Missouri River. Cattle drifted through the streets, gravitating to the livery stable where they could pick up a few wisps of straw.[8]

Weather Bureau records at Bismarck, North Dakota, show that mean temperatures for January and February 1887 were 12.4° and 12.9°F below normal.[9] In log ranch houses and sidehill dugouts ranchers and cowboys tried to block out the bawling of hungry cattle bunched at the corral fences crying for hay that was not available. Ranch employees were found frozen to death near Sundance, Evanston, and Stinking Water, Wyoming. The longing for a warm chinook wind became the yearning for a miracle.

Charles Russell and Jesse Phelps were looking after five thousand head of Kaufman and Stadler cattle. Louis Kaufman wrote Phelps a letter requesting information on how the cattle were surviving. In response, Charles Russell drew his famous *Waiting for the Chinook*. In the drawing, a starved-looking steer stands hunched over in the snow, barely able to stand, while hungry coyotes await the meal soon to be theirs. Russell and Phelps sent the drawing to their bosses without explanation. When it was received in Helena, it caused considerable excitement. Someone added the subtitle "The last of 5,000." This drawing became the symbol of the decline of open ranging on the northern plains.[10]

Below-zero temperatures and harsh winds continued well into the spring season. Cattle cut their legs on the crusted snow, and the scarcity of water contributed to winter losses. Cattle seeking water in the Yellowstone River died when they were pushed into the icy water by the pressure of those behind them. The small creeks and springs of the ranges were frozen. When the spring rains did arrive, they were very heavy, and the weak animals became mired in the mud along the streams.[11]

Old-timers hardened to losses on range operations came near to panic in the spring of 1887. Even the bright young men from the halls of ivy and the drawing rooms of England were shaken by what they saw on the range. A fascinating business had suddenly become distasteful, and many gave up and pulled out.

The extent of the losses in the winter of 1886–87 is difficult to verify. Some numbers were greatly inflated to investors. Some ranchers lost nearly all their animals, especially the recent arrivals. Overall, the Wyoming losses may not have exceeded 15 percent. But the winter caused creditors to lose confidence and forced ranchers to liquidate their starving herds at ruinously low prices.[12]

Estimates of the losses in Montana went as high as $5 million. To meet their debts, ranchers shipped every steer available. Practically none of the animals was in condition for immediate slaughter, and fodder was in short supply in the Corn Belt. There was little demand for feeders. Chicago prices took another drop. John Clay suffered a 25–30 percent death loss on the "moccasin" steers he had bought for Sparks and Tinnin for the 71 Ranch, then suffered the additional blow of having to sell the survivors for less than he had paid. Cattle worth an average of $9.35 per hundredweight on the Chicago market in 1882 brought $1.00 per hundredweight in 1887.[13]

Newspaper editors in the spring of 1887 castigated the giant ranching companies for their "insatiable greed," the cause, in the editors' view, of the winter disaster. Companies financed with foreign capital were especially subject to abuse in the press.

In his memoirs, Granville Stuart was bitter about the treatment he received from the newspapers. He disclosed that while many large ranchers in eastern Montana were losing their own herds, they nevertheless supplied hay to homesteaders so the family milk cow could survive the winter. The farmer-ranchers who had a few cows and sufficient hay benefited by the misfortunes of the large ranchers.

The range in 1887 was nearly devoid of stock as a result of the combination of winterkills and forced sales. The vegetation responded dynamically to the abundant precipitation, and probably to the reduced density of plants caused by drought mortality. The greatly reduced cattle population of 1887 also meant a surplus of cowboys. Unemployment produces many evils, and the wild young cowhands who had been viewed as "knights of the plains" during the previous decade became suspect when calf counts at roundups began to drop alarmingly. Unemployed cowboys-turned-homesteaders sometimes viewed unbranded calves of their former employers as free game.

Much has been written about the economic and social impacts of the drought of 1886 and the hard winter of 1886–87 on the expanding range livestock industry, but the impact on the basic resource of plant communities and supporting soils seems to have been overlooked. John Clay, recognized at the time as an outstanding leader of the industry, spoke of tighter credit as the key to solving the industry's ills; the idea of range management did not even surface. Tight money slowed expansion and may have reduced stocking rates, but it was like giving aspirin to a man with a broken leg.

West of the Rockies, the winter of 1886–87 was rather mild. Did Intermountain ranchers learn from the disaster east of the Rockies? John Sparks had ranching interests in Wyoming, so he was certainly aware of the danger of overextending livestock on the open range without adequate hay reserves.

One of the reasons Sparks moved his operations to Nevada was to take advantage of the Great Basin topography. The altitudes of some of the valleys there were sufficiently low to support desert winter ranges, in contrast to the higher-altitude plains of eastern Wyoming. Also, the drier climate of the valleys appeared to ensure safety from winter disasters. Jasper Harrell had let cattle range freely in that area for a decade with an estimated winter loss of only 1 percent! Again, with hindsight, it seems obvious that Sparks and Tinnin should have considered more carefully the history of the previous two decades. It offered ample evidence of what was to come.

During the disturbances in Utah from 1856 to 1858, for example, a large number of federal troops were stationed in the territory. The freighting firm Russell, Majors, and Waddell contracted with the U.S. government to supply beef to them. Alexander Majors tried to winter thirty-five hundred steers in the Ruby Valley of Nevada during the winter of 1859–60. November brought a heavy snow with cold temperatures. Within forty days all but two hundred steers lay in starved and frozen heaps. The next hard winter was 1861–62. Of a herd of three thousand ewes wintered on the Truckee Meadows, only five hundred survived after more than two feet of crusted snow clogged the valley and attempts to break trail to the Pyramid Lake desert failed.[14]

A colloquialism in the Great Basin is that "only fools and tenderfeet predict the weather." There is a grain of truth in that saying. The timing and quantity of precipitation received in the basin are enormously variable. Tree-ring growth and existing records indicate that rainfall maxima occurred in 1853, 1862, 1864, 1868, and 1890–93. Minima occurred during the 1840s, 1869, 1871, 1889, and 1898. Maxima, or hard winters, came six to seventeen years apart, with most of the intervals at six, seven, or eight years. Intervals between minima, or dry years, were most frequently six to seven years or multiples of these. There was usually a swift succession of dry times and wet times.[15]

The Sierra Nevada blocks winter storm fronts from the Pacific Ocean and casts a rain shadow across much of the Great Basin. Weather on the west slope of the Sierra Nevada also affects the Great Basin, and it was an important factor in the development of the early ranches in Nevada. Dry winters in California usually mean below-average growing season moisture in Nevada. Severe drought in California favored the rapid stocking of Nevada ranges. The green-feed period for the annual ranges of cismontane California, with its Mediterranean-type climate, is October to May. Cattle are dependent on dry feed during the summer drought. If the winter is dry and there is no accumulation of dry feed to carry the stock through the summer, disaster will occur unless other sources of forage are available. During the late nineteenth century, the feed was secured by driving the cattle to the virgin sagebrush/grasslands of the Great Basin. When Nevada ranges were pristine, the consequence of this livestock transfer was a subtle degradation of range plant communities. As the Nevada ranges were exhausted, such transfers were disastrous for both the livestock and the plant communities.

In Idaho and northern Nevada, the winter of 1879–80 was the most severe since 1864. Estimates of losses ranged from 6 percent for Nevada as a whole to 20 percent for northern Nevada and south-central Idaho. Tax rolls indicate a 50 percent decline in cattle herds from 1879 to 1880. Tax rolls are a very poor source for nineteenth-century livestock numbers, but certainly there was an exodus of ranchers from the Great Basin in 1880.[16]

Humboldt Valley ranchers attempted to winter cattle on the Owyhee Desert during 1879–80. There was little snow on the desert floor during most of the winter. Cattle stayed close to existing water supplies because of the lack of water in ephemeral streams. The forage, mainly winterfat and Indian ricegrass, was entirely consumed in the vicinity of the watering points. Extreme cold contributed to losses in the starving cattle. Ironically, while the cattle starved, there was standing forage on the alluvial fans just outside the cattle's ranging distance.[17]

There is evidence that Sparks and Tinnin realized they were nearing or exceeding the grazing capacity of the range during the late 1880s. Newspapers reported that the northeastern Nevada giants of the livestock industry were trying to reduce livestock numbers and upgrade the quality of their stock.[18] The cattle industry in northeastern Nevada had enjoyed good years from 1880 to 1886, and earlier hard winters were forgotten. From 1886 to 1889, precipitation was below normal. Stocking rates exceeded forage supplies during the growing season, and ranchers, including Sparks and Tinnin, found it necessary to ship cattle before they were ready for market.[19]

In March 1889 ranchers in Starr Valley, Elko County, reported ranges in good condition, but by the end of May the Elko newspaper was expressing concern. The growing season for herbaceous vegetation in the northern Great Basin extends from early spring, when temperatures warm sufficiently for plant growth, until the soil moisture is exhausted in early summer. In the spring of 1889, the soil moisture was exhausted in early May on the lower-elevation sagebrush ranges. Perennial grasses withered and lapsed into dormancy. Streams that had been perennial for as long as the oldest settlers could remember shrank to interrupted pools, then dried up completely.

The perennial snowbanks that occupied the glacial cirques on the towering mountain masses that form the headwaters of the Humboldt River shrank and disappeared. For the first time in memory, there was no visible snow on the Ruby Mountains. Hulking like naked giants in the eastern skyline, the Ruby Mountains were a constant reminder to the residents of Elko that they were participants in an environmental event of unusual occurrence.

The Elko papers reported precipitation for the winter of 1888–89 as 3.5 inches. (The mean annual precipitation in Elko for the period 1870–1915 was 9.09 inches.) The weather records for Elko indicate that 6.35 inches of precipitation fell that year; however, this figure includes a large local thunderstorm that dropped 2.80 inches in late May, after the herbaceous range plants had withered and dried.[20] The summer months of 1889 were exceptionally hot and dry, evidenced by dust, withered vegetation, and dried-up streams. The city of Elko was faced with a severe water shortage. The free-roaming horses were concentrated by reduced watering places, and movements of bands could be traced for miles by their dust clouds.

Autumn and Indian summer is usually a glorious period in the Great Basin. Cloudless warm days alternate with crisp nights. Aspen leaves form golden cascades in the draws on the high mountains. The gallery forests along the Truckee, Carson, and Walker Rivers form golden arches over still pools. On October 13, 1889, the first rains broke the drought and settled the dust. The editor of the Elko paper welcomed the rains: "Rainfall this fall is equal to that of all last winter. One foot or more of snow in the mountains last night." There was excitement throughout Nevada. The continuing drought of the previous three years had been broken.[21] Through November, the weather was ideal. Then, on December 5, the white winter struck northern Nevada with full fury, with blizzard conditions for seven consecutive days.

Feed supplies were exhausted by the holidays. Just before Christmas 1889, the season's first snowplow passed through Elko, clearing the Central Pacific tracks of six inches of snow. The paper welcomed the "assurance of a prosperous New Year." But after Christmas the paper began to express concern. Valleys north and south of Wells were "belly deep to a horse." Just before the New Year, eighteen inches fell in Elko on Monday and continued through Wednesday. For the first time since 1862, when the Ruby Valley was settled, Elko County residents did not receive mail deliveries.[22] By early January 1890 snow was two feet deep in the valleys and crusted. January 6, 1890, saw -40°F in Elko, warming six days later to -36°F. Houses creaked in the night as contracting timbers pulled at square-cut nails. Snow north of Elko was reported to be forty-two inches deep, and the stage to Tuscarora was stranded.

The stage to Twin Falls was the communications link for the Sparks-Tinnin ranches. The Salmon Falls Creek area around San Jacinto, a major Sparks-Tinnin ranch, is the coldest part of Nevada, with -50°F recorded and average annual snowfall of 28.6 inches.

The January 12 edition of the Elko paper reported that ranchers were planning to ship their cattle to California, but this plan received a setback when the tracks were blocked both east and west of Elko. That same week, the Elko editor found a half inch of ice on the water in his well.[23]

In mid-January northern Nevada received six more inches of snow. Halleck Cattle Company cowboys wore all the clothes they owned as they pushed their starving cattle through frozen willows along the Humboldt River to isolated patches of feed. Sloughs piled up with dead cattle. At a

Humboldt bridge, thirty-nine cattle were caught at one time. On January 15 the Elko paper reported that several thermometers had registered near -60°F. The official reading was -42°F.

Low temperatures continued through February, from -40°F on the first day to -41°F recorded on the last day of the month, with additional new snow. The Nevada Land and Cattle Company estimated that its winter losses had already reached 98 percent. Stored hay could not be moved over the drifted roads. In Secret Valley, Elko County, A. G. Dawley reported fifteen-foot drifts between his house and barn, where he carried feed and water through a tunnel to save two stallions.

There is a love-hate relationship between herdsmen and their domesticated charges. They take pride in their animals from birth to maturity, then ship them off to slaughter. Only the most calloused rancher could stand by unmoved while his animals suffered. Small homesteaders with twenty cows knew each animal well. Families had scraped, saved, and done without to accumulate them. They represented the means of paying off supplies bought on credit, that rare new dress for overworked wives, and toys for the children next Christmas.

On those -40°F, -50°F, or -60°F nights, homesteaders herded their animals around bonfires and fed handfuls of native-grass hay, cut the previous summer by hand, to the weakest animals. In this area where structural timbers and lumber had to be shipped great distances, houses, sheds, and outbuildings were built from native stone. Most buildings were roofed with juniper or aspen rafters, thatched with woven willow sticks and Great Basin wildrye stems, and covered with dirt. In the desperate white winter of 1889–90, ranchers stripped their roofs to salvage feed to keep a saddle horse alive a few more days.

By late January things were very bleak throughout the Intermountain area. Elko townspeople shoveled snow from roofs and canceled church. Finally, in desperation, cattle turned to browsing the sagebrush and died of the malady Nevada ranchers call "hollow belly." Sagebrush inhibits the activity of rumen microorganisms. An autopsy of an animal forced to eat it reveals a rumen packed with sagebrush rich in nutrients but indigestible. The rumen is vital to the winter survival of ruminants. The rumen microflora create heat when they break down high-cellulose-content forage. A cow with a full rumen can withstand bitter cold. An empty rumen means a cold cow, soon dead.

The question was no longer whether many cattle would die, but how. One herd of 300 broke into a stockyard, and 117 smothered in the crush to reach the hay. Twenty-six cows crowded into a Starr Valley cave to escape the storm; all perished. Horses bunched up to chew each other's manes and tails until all the hair was gone, then died in a group. Animals marched up and down seeking the herdsmen to which their ancestors, millennia ago, had given bondage in return for care and subsistence. The herdsmen were not able to fulfill their responsibility, and animals died by the thousands.

Late in January, L. A. Nelson, in charge of Sparks-Tinnin cattle on the Salmon River, reported in the Elko paper, "We have a foot of snow on the winter range. Cattle have been shrinking very much the last 10 days. Unless we have a thaw in a short time, there will be a good many of the old cows that will turn up their toes before spring, as there is no hay to feed anything but saddle horses." On the North Fork of the Humboldt and in Independence Valley, even sheltered cattle on feed were dying from the cold.[24]

In mid-February the editor of the Elko paper interviewed John Sparks as he passed through Elko returning from an inspection of his ranches. Sparks cautioned the editor to keep a stiff upper lip and to avoid printing scare stories. He said his firm had lost more stock on two previous occasions than they had that winter. He had "traveled to the Snake River and back looking for dead cattle and found few except for thin and weak stock." He claimed to be feeding eighteen thousand head.[25] This is surprising considering his foreman Nelson's report in late January that they were out of hay. The editor concluded that Mr. Sparks was not guessing, had seen the range, and understood range country. Sparks repeated his story when he reached Reno, saying that "he had ridden 400 miles on horseback without seeing many dead cattle."[26] On February 10 the same paper had carried a story reporting that N. H. A. Mason had spent two months in the snow in a futile attempt to save his herds. The winter broke Mason, and his interests were absorbed by Miller and Lux.

February 9 saw the first train reach Elko from San Francisco since January 15, when snowfall in the Sierra Nevada had blocked the tracks. The city of Elko used railcars to haul snow out of town.

On March 2, 1890, several thermometers registered -40°F just before moderate temperatures arrived to turn the deep snow to slush. March 17 is a day that stands out even in that formidable winter. Storms over southern Idaho and northern Nevada began with rain and sleet, then turned to snow. Drifts of heavy slush on the lower sagebrush ranges refroze with icy-hard crusts. Rain and sleet saturated the shaggy winter coats of livestock that had survived and were searching for forage on the open range. The temperature dropped, and the weakened animals were unable to shake the ice from their coats.[27] The Nimmo Report lists late spring storms as a separate category of death losses for the range livestock industry.

The white winter of 1889–90 was severe in virtually all the states and territories west of the Rockies. The Pacific Northwest had one of the four worst winters recorded before 1900, and the governor of Idaho labeled it "the most severe winter ever experienced since settlement of this country." Similar reports came from the Big Bend of the Columbia to Lake County in south-central Oregon, and from the Yakima Valley to the Snake River Plains. The same picture came from the Palouse country southward through the Walla Walla Valley and northeastern Oregon to Owyhee and Malheur Counties. Taken together, the reports constructed a gruesome pattern of cattle dying from lack of feed, water, and shelter.[28] The valleys of central California became lakes during the classic winter of 1889–90. Henry Miller spent the entire winter stemming one disaster and then rushing to cope with the next.[29]

The L-7 Ranch near Baggs, in western Wyoming, was owned by one of the Swan brothers, a family closely associated with the winter disasters of 1886 and 1887. L-7 cowboys drove their cattle onto the Red Desert north of Rawlings to let them fend for themselves on the desert shrubs. On the day before Christmas 1889, they turned loose ten thousand cows. Their losses were estimated at 75 percent of the cattle and 66 percent of the saddle horses.[30] One newspaper editor wrote that the ranchers of the far West could take consolation from the fact that "they all went down together."

The regional extent of the disastrous winter of 1889–90 made it difficult to restock the ranges in 1890. The trail drives from Texas had been shut down by the settlement of western Kansas and the Texas fever controversy. Nature's horrors fade with time, however, and in April 1890 spring finally came to the sagebrush/grasslands. Riders covered their noses with bandannas to stifle the stench of carcasses thawing in the retreating snowdrifts.[31] East of Elko, one drift revealed five cows, one horse, two mule deer, and one pronghorn huddled together in death.

It was said that a man could walk on dead cattle for one hundred miles along the Marys River fork of the Humboldt. J. D. Bradley of Mason and Bradley reported that the country around Deeth and Halleck was strewn with dead cattle. He told the Elko paper that the estimated loss of $750,000 in stock for the country was entirely too low. J. R. Bradley reported counting four hundred carcasses of dead cows within four hundred yards.[32] A. G. Dawley from Halleck expressed fear that the winter would end the cattle business in Ruby Valley.[33] A rancher wrote from Ruby Valley to the Elko paper, "I've just come in from four days riding—counted 100 dead cattle along the roads." Horses that survived had hair and skin worn off their legs and noses from rooting for feed.

Carrion feeders waxed fat and happy. Cattle and horse carcasses were found wedged in the top of juniper trees where they had walked out on the top of snowdrifts. When the ice thawed, the streams were choked with carcasses. In the Humboldt River, dead cattle jammed against the bridges. Downstream towns complained of the stench. People who depended on the streams for domestic water had to quickly sink wells to save themselves from a greater epidemic. In the spring of 1890, a district judge went to hold court at Challis, Idaho. He found the stench through part of Lost River valley so bad that he issued a court order to county officers to get the carcasses burned and buried.

Estimates published in the Elko papers reported that some large ranches had lost 95 percent of their livestock. Sparks-Tinnin had branded thirty-eight thousand calves during the 1885 roundup on their Nevada and Idaho holdings; in 1890 they branded sixty-eight calves.[34] John Sparks later described his losses to a reporter for *Harper's Weekly*. "In 1889 and 1890, Mr. John Tinnin and myself were ranging in Elko County, Nevada, and Cassia County, Idaho, 65,000 head of cattle—we lost that winter, which was a severe one, 35,000 head of cattle." Sparks made this comment in 1902, twelve years after the white winter.[35]

After the spring roundup in 1890, John Sparks, Jasper Harrell, Andrew J. Harrell, and John Tinnin met in a cabin on Cottonwood Creek south of Twin Falls, Idaho. John Tinnin, the Confederate colonel with the steamboat Gothic mansion in Georgetown, Texas, was broke. Jasper Harrell had sufficient capital and confidence to continue in the livestock business. Sparks must have had adequate resources as well because he handed Tinnin forty-five thousand dollars in cash for his interest and suggested to Tinnin that he go to the sandhills of Nebraska to start again. He promised to send Tinnin cattle to stock his range.[36]

After this meeting a new company was incorporated in California: the Sparks-Harrell Company, with a capital stock value of one million dollars. According to the incorporation papers, the stock was fully subscribed, with John Sparks subscribing to 50 percent, Jasper Harrell 49 percent, Andrew Harrell 0.8 percent, and Martha and Ella Harrell 0.1 percent each. The Harrells listed their home address as Visalia, California; Sparks listed his as Georgetown, Williamson County, Texas.[37]

*   *   *

The transplanted Spanish system of open-range livestock was dead. Many of the stockmen who had brought Spanish Longhorns from California or Texas to the sagebrush/grasslands had been wiped out. The Spanish vaqueros from Texas or California and the cowboys from Texas drifted away. The Spanish left behind a sprinkling of place-names and a rich vocabulary of technical terms concerning horses, riding equipment, and the handling of livestock on the open range. Most important, the vaquero had imparted the basic skills necessary to work with cattle and horses on the open range.

The loss of cattle and the vacant ranges that resulted had a lasting influence on the livestock industry. Critical conflicts surfaced between stockmen and sheep men. The range sheep industry had fewer losses because sheep were better adapted to the environment and forage base of the desert ranges. The range sheep industry was also smaller, with fewer marginal operations. After 1900, when the range sheep industry peaked, there were many examples of excessive winter losses. But the net immediate effect of the white winter was freedom for the range sheep industry to expand without competition from previously established cattle ranches.[38]

The superabundant precipitation of the hard white winter promoted excellent plant growth during 1890. The ranges were virtually empty. The pristine plant communities of the sagebrush/grasslands had been severely reduced by two decades of unlimited livestock grazing. Domestic livestock had selectively exploited the perennial grasses, leaving the shrubs to take advantage of the near vacuum in the spring of 1890. Shrub establishment included stands of the desirable browse species bitterbrush, but it also included an overwhelming abundance of toxic big sagebrush and brought about a basic change in the forage resources of the sagebrush/grasslands.

Pinyon/juniper woodlands greatly expanded their ranges during the decade of the 1890s. Many of the woodlands in Nevada had been severely depleted during the 1870s and 1880s as energy sources for the mining industry.[39] The combination of three good precipitation years at the start of the decade also favored woody vegetation.

The grasslands of the high plains responded dynamically after the drought of 1886 and the abundant precipitation of the winter of 1886–87. Many of the dominant grasses of the high plains grasslands are rhizomatous, and they responded vegetatively to occupy the environmental potential released by drought losses once the rains returned. Great Basin perennial grasses are largely bunchgrasses that depend on seed for reestablishment. Two decades of excessive grazing had virtually eliminated seed production in many areas. Without seed reserves in the soil, there was no way for the grasses to respond to the abundant precipitation of the winter of 1889–90. After the reduction in livestock numbers caused by the winterkill, the perennial grasses doubtless produced abundant seed crops in 1890, but shrubs, pinyons, and junipers had the advantage of earlier establishment and preempted the released environmental potential.

The great stillness on the ranges in the spring of 1890 was broken during the summer and fall by the sounds of the pack mules and wagons of bone pickers, a profession born in the wake of the near extinction of the American bison on the Great Plains. The bone pickers built white hills of bones along the railroad sidings at Montello, Toana, and Wells after the white winter. The remains of the great expectations of the early livestock men of the sagebrush/grasslands were boiled for oil, ground for fertilizer, or cut into buttons by the Pacific Fertilizer Company on the shores of San Francisco Bay.

CHAPTER 8

Water
The Finite Resource

The winter of 1889–90 drove home the lesson that forage had to be conserved for wintering cattle on most of the sagebrush ranges of the Intermountain area. But where was the necessary hay to be produced?
Most of the native plant communities were not suited for harvesting hay.
The density of grasses and the total biomass of herbage they produced
were not suitable for hay production, and they contained shrubs that
were too woody for mowing. Forage crops for hay production in the sage-
brush/grasslands will not grow without irrigation, but only 2 to 5 percent of northern Nevada is irrigable with water diverted from stream
flow. The tiny portion of the land that could be irrigated became the
controlling element for vast areas of rangeland.

Irrigation was the only way to produce forage in sufficient quantity to
make harvesting hay economically feasible. Early ranchers such as Jasper Harrell had nature as a model to follow. Each spring when the snowmelt occurred, streams such as Goose Creek overflowed their banks and
flooded the natural meadows, including Harrell's Winecup field.[1]

Irrigation is such a commonplace in the western United States that it
is difficult to comprehend that it was not always a part of American agri-

culture. The Mormons who settled Salt Lake Valley are generally credited with being the first Anglo-Americans to raise crops under irrigation
in the western United States. On July 25, 1847, the day after the first small
group of Mormon pioneers arrived in Salt Lake Valley, a special service
of thanksgiving was held. Orson Pratt was the principal speaker. He developed a prophetic vision of the parched desert land on which he stood.
His text was from the words of Isaiah: "The wilderness and the solitary
place shall be glad for them, and the desert shall rejoice and blossom as
the rose."[2]

The Mormons were not the first to raise irrigated crops in the western United States, of course. Sophisticated Indian cultures of the Southwest had developed irrigation systems that flourished and disappeared
centuries before the Mormons arrived in Salt Lake Valley. Among the
tribes of the Great Basin, Shoshone family units sometimes diverted
ephemeral streams to increase grass seed production.[3]

Although the Mormon settlers are often credited with the conceptual
model of modern irrigation in the far West, a second center of development was the gold-mining industry of California. Virtually all forms of
placer mining required the diversion of water. Hydraulic mining involved the diversion and long-distance transfer of sufficient water to
power the hydraulic giants. Technology that was developed for diversion
dams, flumes, and inverse siphons could be transferred to irrigation
works in other areas. In California, many of the water diversion works
that were developed and built for mining were converted into irrigation
works. This technology was spread to other territories by miners. David
Wall, who gained experience in the California goldfields, is often credited with developing irrigation systems in the Platte River valley of eastern Colorado independent of any Mormon influence.[4]

The structures and agronomic principles needed to raise crops under
irrigation were not the only things lacking in the Intermountain area;
laws governing water rights were needed as well. The Mormon settlers
had the structure of a church-oriented society to govern their initial

water policy. In the early Mormon settlements, each head of a family was given 1.25 acres in town. Those who followed occupations in town were given 5-acre garden plots. Farmlands were assigned in 10- to 80-acre tracts depending on their distance from town. In order to raise crops, the Mormons had to dig irrigation ditches to water each plot. The ditches were owned in common.[5]

In the rest of the West, water laws generally evolved from mining laws. In the eastern United States during the early nineteenth century, diverting water to power mills was recognized as a legal use of the resource. After water falling over the mill wheel had provided the energy to run the mill machinery, it was returned to the stream in virtually the same amount that was diverted. In the West, however, water used in mining or in irrigation seldom returned to its natural channel undiminished in quantity.

Water laws in all states of the United States are based on either the riparian doctrine or the appropriation doctrine. The riparian principle is the right to use water, which is a real property right for most purposes, based on the ownership of land next to or contiguous with surface water. Under the common-law riparian doctrine, the right to use water is inseparably annexed to the soil. Use does not create the right, and misuse does not destroy or suspend the right. All states east of the ninety-eighth meridian except Mississippi and Florida follow the riparian doctrine.[6]

All states west of the ninety-eighth meridian, as well as Mississippi and Florida, recognize the appropriation principle, which bases water rights on beneficial use of the water. The first to appropriate and use the water has a right superior to the rights of later appropriators.

Nevada and the other western semiarid states were granted control of their natural waters for appropriation by citizens for irrigation and stock water by the federal government. Legal difficulties over water use started with the discovery of the Comstock Lode. The waters of the Carson River provided power for stamp mills for ore reduction in the Dayton area, transportation for wood products from the forests of the Sierra Nevada to the mines and mills, and irrigation water for the ranchers of the Carson Valley. The conflicts and resulting litigation that developed among these users is well documented in the publications of Grace Dangberg and John Townley.

The 1889 Nevada legislature provided a means for determining individual water rights. The act, designed to regulate the use of water for irrigation and for other purposes, was modeled on Colorado law and imposed a self-regulating system for determining water rights. The law divided the state into seven irrigation districts by major drainage basins and created water commissioners for each district with the authority to decide individual water entitlements within their districts. The law further decreed that all rights to water were to be filed with each county recorder by September 1, 1889; reserved unappropriated water to the state; and prevented enlargement or erection of irrigation works without express permission from the appropriate water commissioners. After the law was passed, landowners began to file their claims to irrigation water with the various recorders. Individual claims were commonly exaggerated and far exceeded the ability of most streams to supply.[7]

This first major attempt by the Nevada legislature to regulate water for irrigation had a short and rocky life. The years 1888 and 1889 were extreme drought years in the Great Basin. The crucial legal conflict took place in Humboldt County, where Judge A. F. Fitzgerald heard the suit entered by P. N. Marker et al. against some 540 Humboldt River valley irrigators. Drought conditions during 1889 forced Lovelock farmers to bring a common action asking that all Humboldt rights be determined and enforced. Upstream ranchers in Elko County argued that the court could not determine individual water entitlements on the Humboldt because the 1889 water-control statute was unconstitutional. Judge Fitzgerald agreed, so the basic point at issue, riparian versus prior appropriation rights, was never considered. The 1893 Nevada legislature repealed the now-defunct 1889 water law but failed to enact anything in its place. The Nevada legislature next passed laws establishing the procedures for registering water rights in 1905. Water already appropriated for beneficial use at the time of the act was recognized as vested rights. The act provided that future appropriations were to be made by the state engineer. The water-rights law in Nevada did not specify how long people with vested rights had to record these rights.[8] Strange as it may seem, some ranchers, whose very livelihood depended on irrigation, delayed filing on vested rights that dated from the 1870s until the 1920s and early 1930s. The problem with the delayed filing on vested rights was that the dates, locations, and amounts of original appropriations were based on the memory of individuals who appropriated or witnessed the appropriation of water as much as fifty or sixty years prior to the filing.

Nevada, along with seven other mountain states, follows the appropriation doctrine exclusively. This doctrine is well suited to areas that require the consumptive use of water; that is, use that takes a substantially larger quantity of water out of the stream system than is returned after the water is used. The owner of an appropriation water right is entitled to use a stated amount of water even if such a use means that other would-be users with later priority rights are deprived of water. Most states date an appropriation water right from the first date water was put to a beneficial use, counting the date construction began on the diversion works as the first use.

According to federal court records, the Winecup fields of Jasper Harrell were first artificially irrigated on May 1, 1875, and at that time only 100 acres were irrigated. In 1886 irrigation was extended to a additional 125 acres, and in 1900 an additional 100 acres were irrigated—this on a ranch with thousands of acres of sagebrush rangeland. The Rancho Grande, Sparks's favorite ranch in the Intermountain area, had a similar sequence of irrigation development, with 300 acres developed in 1883 and 150 acres added by the turn of the century.

The Salmon River, or Salmon Falls Creek, originates and extends for many miles in Nevada, running in a generally northerly direction until it crosses into Idaho and discharges into the Snake River. The drainage system belongs to the Snake River hydrographic basin rather than the Great Basin. The stream had to fight to escape the confines of the Great Basin. North of the Vineyard and Hubbard Ranches, the intrusive granite rocks that support the mineralized area around Contact were almost too indurate for the channel to wear through; again, north of San Jacinto, basalt flows reach down from the highlands and threaten the course of the river. The major portion of the Salmon River's water is supplied by O'Neil Basin and Shoshone Creek. The headwaters of O'Neil Basin are the ten-thousand-plus-foot peaks of the Jarbidge area. The river leaves O'Neil Basin through a narrow gorge and joins with Jakes Creek, which flows in from the south at the Vineyard Ranch. The river flows in a northerly direction from the Vineyard Ranch to the state line and is joined by Trout Creek and Shoshone Creek a few miles south of the state line. Shoshone Creek, a principal tributary, rises in Idaho and flows south, crossing the state line east of present Jackpot, Nevada. It then flows in a generally westerly direction and joins the main stream at the lower end of the San Jacinto Meadows. The Salmon River was by far the largest stream in Sparks's ranching empire and the major base of irrigated hay production. Eventually more than ten thousand acres of irrigated land was developed along the Salmon by Jasper Harrell, Sparks-Tinnin, and Sparks-Harrell.[9]

According to the district court records, Jasper Harrell started irrigation works on the Salmon River in 1873. The first irrigation season was 1874. The East Boar's Nest Slough Ditch, the West Boar's Nest Ditch, the Salmon River Sloughs, and the Harrell or Big Ditch brought more than two thousand acres under irrigation. In 1878 another eight hundred acres were brought under irrigation, mainly on tributary streams such as Jakes Creek to the south of the Vineyard Ranch. For the next ten years the irrigation system remained relatively stable while the number of livestock run on the ranges expanded by at least a hundredfold. Sparks and Tinnin were not very active in enlarging the Salmon River irrigation system. In 1887 the Willow Spring or Dry Creek Ditch brought about seven hundred more acres under irrigation, but this was the only project developed while John Tinnin was Sparks's partner.

In the wake of the winter of 1889–90, the need for additional hay production was obvious. When Jasper and Andrew Harrell returned to the Sparks Company, they started new, higher-elevation ditches and extended the existing ditches. By 1894 about ten thousand acres had been brought under irrigation. A major part of this addition was the Highline Ditch, which added twenty-five hundred acres.[10] By 1900 Sparks maintained that his network of ranches produced 15,000 tons of hay per year, but more important, his Great Basin wildrye meadows produced the equivalent of 100,000 tons of standing cured hay. This would indicate that Sparks and Harrell had fenced fifty thousand acres of naturally occurring Great Basin wildrye communities. John Sparks enjoyed bragging about his meadows on Salmon Falls Creek. His idea of a suitable photograph of his ranch was a picture of him in his fancy buggy with a flashy team of horses driving through Great Basin wildrye higher than the buggy wheels.

In the diversion of irrigation waters, ranchers followed the procedures developed by placer miners. A rock dam—usually a pile of rocks braced with cottonwood, aspen, or any other available timber—was built across the stream at the point of diversion. Often the first spring flood would wash out the dam, and rock would have to be hauled in wagons to repair the dam before irrigation could start.

The first ditches had their origin on the land claims to be irrigated. Water was diverted on the upstream end of the claim and spread as far as the grade would allow. Most early land claims were on streams in valleys. Simply by diverting water on their own land, claim owners could irrigate the lowlands along the streams. To get water up to the first stream terrace above the bottomlands it was necessary to put the point of diversion further upstream to maintain the necessary elevation of the ditch.

In the Bruneau River valley of southern Idaho, John Baker dug the first ditch in 1876. Each rancher, large and small, with land along the Bruneau

River dug ditches throughout the late 1870s and early 1880s. It soon became apparent that much longer and larger ditches would be needed to bring water to the upper terraces. The soil of the first terrace above the stream bottoms was usually deep and finely textured, and the great size of the sagebrush growing in it indicated its potential productivity. When it became obvious that the necessary irrigation works were on a larger scale than individual settlers could afford to build, individuals joined together to form irrigation companies. Many such companies were formed throughout the West in the 1880s. They varied in their design and complexity. Some were informal arrangements among groups of ranchers who shared work on a ditch. Some companies sold shares of stock to obtain money with which to build the dams and ditches necessary to make the system work.[11] The acquisition of water rights and the development of irrigation systems were among the stated purposes of the incorporation papers of the Sparks-Harrell Company. The company undertook its own water development.

Building an irrigation system was difficult and time-consuming. Teams drawing Fresno scrapers helped construct ditches through alluvium; rocky slopes and cuts were pick, shovel, and blasting operations. Construction of ditches and irrigation structures was largely by hand labor. The cost of digging the Orr Ditch from the Truckee River varied from $0.75 to $10.00 per rod (sixteen feet) depending on the nature of the ground through which the ditch was being constructed.[12]

Often, irrigation water was simply flooded out onto uncleared land. The native desert shrubs could not stand the flooding and soon died. Cattle wintering on the newly irrigated lands trampled the brush into the soil and it soon disappeared. The first terrace often contained a few wet meadow species scattered under the shrubs. The native sedges and rushes rapidly increased on the newly irrigated soils, and soon the first-terrace sites supported plant communities remarkably similar to those on the floodplain.[13]

Many ranchers attempted to seed their newly irrigated lands. A familiar sight was a mounted rancher crossing his new field with his winter felt hat pulled down hard against the March winds as he dipped into a gunnysack of grass seed hooked over the saddle horn and broadcast the seeds to the winds. Favored grass species included orchard grass, timothy, and redtop. Many of the early ranchers were familiar with agriculture in humid environments, and these species fit their conception of what a wet meadow hayfield should contain. All of these species naturalized in the Intermountain area, but they rarely dominated what became known as the native hayfields.

The land usually was not leveled before irrigation. Natural drainage ways and swales gradually sodded in and lost some of their relief. Irrigating these undulating meadows became an art. In the spring, wagonloads of manure were hauled to the fields to make strategically located dams. Manure from work and saddle horses that were kept inside during the winter provided a convenient construction material. Tremendous heads of water were used in the spring. The ranchers called it "taking the frost out of the ground." The low places might be four feet deep in irrigation water, and the high places took care of themselves.

Old-time ranches that obtain their irrigation waters from cooperative farmers' ditches are often advertised as having "free" irrigation water, as opposed to land in formal irrigation districts, where a fee is charged for the water. The "free" water is very much a misnomer because the ranch owner has to maintain the ditches and diversion structures.

Mucking ditches in the spring became a familiar part of western ranching. Standing in gummed rubber boots in six inches of mud in the bottom of an irrigation ditch while cutting and lifting heavy sod out of the ditch was a long way from the "knight of the plains" concept of a cowboy. The mucking was a spring social event when ranchers worked together on cooperative ditches. Ranch jobs in which crews had the time, were close enough together, and it was quiet enough for conversation were few on nineteenth-century ranches. Mucking ditches was one of the rare jobs that allowed the crews to visit. The event usually ended in the stock argument of the bunkhouse crew, which pitted the tough, with-drawn old-timers against the brash younger workers.

Cutting sod requires a sharp shovel, and the conversation of men toiling in the muck was punctuated by the harsh ring of a fine flat file milling the cold steel of a good shovel blade to a razor edge. More than one cowboy-cum-ditch-mucker overestimated the weight of sod he could pitch out of a ditch and ended up with a hernia.

The newly created native hay meadows were generally irrigated—or more correctly, flooded—only once each season. When the low spot had more or less dried out, it was time to mow the forage for hay. On exceptionally wet years, the bottomlands sometimes failed to dry sufficiently to allow the hay to be cut and cured. Delays caused by high water were especially common in fields along the Humboldt River.

The practice of the early ranchers, who had the oldest-priority water rights, of raising native meadow hay under flood irrigation came under severe criticism from agriculturists and other would-be waters users. The waters of a given stream were appropriated by successive diversions until all possible water was placed in use. The climate of the Great Basin is highly variable. There is really no such thing as an average year in terms of an actual pattern or amount of precipitation. Averages are products of the statistical treatment of extremes, and droughts are an all too common part of the Great Basin environment. When droughts occurred, the ranchers with the oldest water rights diverted water as usual and flooded their meadows; ranches with later rights often had to do without irrigation water.

The Nimmo Report describes the water conflicts prevalent in the early 1880s:

A certain number of farmers unite and build a ditch from a river to irrigate their lands; an adjoining set build another for their lands, and so on until several ditches are in operation. The ditches being under separate management, conflicts ensue and damage suits follow, as one company fails to take care of its waste water, allowing it to run on the farm of a stockholder of another company, thus wasting water that might be used to an advantage in another direction.[14]

Water rights, like all property rights, are subject to qualification and to regulation by the state. There is no such thing as absolute ownership. Society relies on the law to settle conflicts over water, but the concept of wasteful use has rarely been introduced to water law to enforce efficient irrigation practices. Courts have thus been faced with the question of how to increase efficiency without infringing on the basic property right connected to water appropriations for irrigation. The failure to increase efficiency in irrigation had social ramifications throughout the Intermountain area. Technologically advanced agronomic farming was restricted because the old-time ranches with priority to water rights continued the wasteful practice of flooding the native meadows for hay production. This kept northern Nevada in a cow-country economy.

Miles of ditches were required to irrigate lands farther from and higher in elevation than the floodplain bottoms. The diversion point was usually a long distance from the area to be irrigated. The Nimmo Report indicates that in 1884 there were eight hundred irrigation ditches in Nevada, aggregating about two thousand miles in length and irrigating 150,000 acres. Obviously, many ranchers were dependent on lifeline ditches strung out for several miles.

The irrigation had some unexpected results. Soils of arid regions develop under low rainfall and scant vegetation, and are different from soils of humid areas. Arid soils have been weathered less and contain smaller quantities of organic matter than their humid-region analogues. Desert soils usually contain appreciable quantities of lime, magnesium, and other mineral elements in unweathered and precipitated minerals. Such soils are normally alkaline in reaction and in some situations are enriched in basic minerals through irrigation.[15] These distinctive characteristics of arid soils must be considered in managing them for crop and forage production. While humid-soil farmers have problems with acidity and the need to add lime, farmers on arid-land soils must deal with excess lime, alkalinity, and soluble salts. Irrigation increased these problems.

The Mormon pioneers who arrived in Salt Lake Valley on July 24, 1847, diverted the waters of City Creek for irrigation in order to plow the dry, baked land. When this water was put on land that had developed under a mere fifteen inches of rainfall per year, a new cycle of nature began. The fifteen-inch annual rainfall had fallen on soil whose potential evaporation was sixty inches. Consequently, leaching had occurred only during exceptionally stormy periods, and these infrequently percolating waters had not reached any but the highly soluble products of rock weathering from the soil. Suddenly, with irrigation, the natural balance between precipitation and evaporation was reversed.

The first unfavorable result of irrigation noted was the gradual buildup of groundwater in some areas. The first lands to be irrigated were low-lying lands of fairly level topography. Later, when water was diverted onto higher bench lands, the drainage water seeped down into the land below, reducing aeration and bringing in excess salts until, in many cases, these once highly productive lands were reduced to low production levels. The proper balance among water, soil, and crops is difficult to determine and maintain. Excessive use of water leaches many nutrients required by plants from the soil and leads to problems with drainage, aeration, and salts. Economical use of water leads to deposits of undesirable minerals in the soils irrigated.

Irrigation waters usually contain larger amounts of dissolved impurities than natural precipitation does. Sodium is the most common injurious ingredient of irrigation waters. Sodium is no more toxic to plants than calcium or any other element, but if the concentration of sodium exceeds that of calcium and magnesium, the sodium will adsorb to the clay particles, displacing the normal high proportion of calcium. This causes the clays to disperse and lose their normal structure. Farmers call such soils "slick spots." They have virtually no drainage and are very poor for plant growth. As early as 1880, farmers in Nevada were experimenting with the addition of gypsum to counteract saline/alkaline soils.[16]

Ranchers and irrigators took all the water they could get and all that their ditches would hold. Frequently, the waters of given streams were completely appropriated before water users with lower priority got to use them. Thousand Springs Creek Valley, for example, was a valuable part of the Sparks-Harrell ranching empire. There are approximately fifteen thousand acres in the valley that can be readily irrigated with water diverted from the streams or the many springs. The available water averages eighteen thousand acre-feet annually. (An acre-foot of water is the amount necessary to cover one acre of land to a depth of one foot.) Most of the irrigable land occurs on the branches of the stream near Montello, Nevada, where Sparks's Gamble Ranch was located.[17]

Snowmelt in the high mountains of the Thousand Springs Creek watershed produces maximum seasonal stream flow during April, May, and June. Low water occurs from August through October. The valley floor of Thousand Springs Creek receives five to eight inches of annual precipitation. The mountains that form the headwaters of the stream receive fifteen inches of precipitation. Irrigation is totally dependent on the runoff from the high-elevation precipitation. The condition and quality of the watershed eventually determine the tenure, quality, and quantity of irrigation developments. The quantity and density of plant cover on the soils of the watershed contribute to the rate of runoff and the amount of siltation as soils are eroded. Destruction of the vegetation cover by overgrazing eventually leads to floods. During the period from 1900 to 1920, tremendous floods came down Thousand Springs Creek to pass through Montello and out onto the salt flats of pluvial Lake Bonneville. These disastrous floods occurred well after the time scale of this narrative, but their seeds were planted in the grazing practices of the 1870s.

The need for a federal land policy that would encourage and accommodate irrigation of western lands became obvious in the 1870s. The Desert Land Entry Act of March 3, 1877, allowed entry on 640 acres of land, four times the amount of land allowed under the preemption or homestead methods of land entry. Eastern congressmen, against large-scale giveaways of public land, fought long and hard to block the passage of this act, while representatives of western states insisted that the cost of developing irrigation systems made the larger acreage necessary. After many complaints about illegal land entry under the Desert Land Entry Act procedure, the law was amended in 1890 to make 320 acres the maximum that could be obtained.[18]

The original law of 1877 was restricted to the states of California, Nevada, and Oregon, and the territories of Arizona, Dakota, Idaho, Montana, New Mexico, Washington, Utah, and Wyoming. It was amended in 1891 to include Colorado. The stated purpose of the Desert Land Entry Act was "to encourage and promote the reclamation, by irrigation, of the arid, and semiarid lands of the western states." It was assumed that settlement and occupation would naturally follow when the lands had been rendered more productive and habitable.

To be eligible for entry under the act, the land had to be irrigable and not capable of producing a crop without supplemental irrigation. An exception to the crop rule was alternate-year grain/fallow systems farming. Application for entry had to be accompanied by evidence of water rights already acquired by appropriation, contract, or purchase of a right to permanent use of sufficient water to irrigate or reclaim all of the irrigable portions of the entry. Entries had to be as nearly square as possible even when they involved unsurveyed land. This specification was a major problem when irrigable acres were restricted to long, narrow strips along streams. The entire entry had to be in one piece and could not be in disjunct tracts.

There was probably more fraud connected with Desert Land Entry Act entries than with any other method of disposal of the public domain during the late nineteenth century. In this connection, it is worthwhile to note the procedures used for establishing the final proof for such entries. The claimant had to name four witnesses for the final examination who were familiar, from personal observations, with the land in question and what had been done toward reclaiming it. The examiner chose two of the four witnesses to interview.

The final proof had to show specifically the source and volume of the water supply—the number, length, and capacity of ditches on each of the legal subdivisions. Witnesses had to state that they had seen the land irrigated and the different dates on which they saw it irrigated. As a general rule, actual tillage of one-eighth of the land had to be shown. The original law had required that all 640 acres had to be irrigated, but this was practically impossible. It was not sufficient to show a marked increase in forage production from native grasses with irrigation. An actual crop had to be planted and irrigated.

Despite the fact that the idea for the Desert Land Entry Act originated with experiments in Lassen County, California, and was specifically designed for Nevada, this method of disposal of public lands was never popular in the Great Basin. In Nevada, 611,320 acres were entered under this act and 143,000 were finalized. There was plenty of irrigable land in the Great Basin that would respond to the application of water. Unfortunately, there was not sufficient water to irrigate more than a small fraction of this available arable land.[19]

The development of irrigation in the sagebrush/grasslands environment in the nineteenth century had three highly significant results. First, the available forage base was greatly increased, and this forage could be conserved for winter hay feeding. Second, the successful production of irrigated crops in the Intermountain environment provided the germ of the idea that developed into the government-sponsored land reclamation projects of the early twentieth century. Third, the incorporation of irrigated farming with range livestock operations contributed to the further grounding of the cowboys. The wild horseman of the plains lost considerable glamour but became more functional by standing in a muddy ditch leaning on a shovel.

Part III The Land in Transition

The cold deserts are lands of extremes. Bitter cold and snow are followed by burning heat and drought. The environment of the Great Basin had to be modified to permit the raising of cattle on a sustained basis. Such modifications included the painfully slow digging of irrigation ditches to bring the land into production for crops of hay. To come to the sagebrush/grasslands and raise cattle, ranchers had to be brave, and to survive in such enterprises they had to learn that water runs downhill—most of the time.

CHAPTER 9

Making Hay in the Great Basin

A cowboy in the 1890s lamented, "Cowboys don't have as soft a time as they did. I remember when we sat around the fire the winter through and didn't do a lick of work for five or six months of the year except to chop wood to keep us warm."[1] With the advent of hay production, it was irrigation and haying in the summer, and feeding the darn stuff in the winter. All of the old riding, roping, and branding was sandwiched into the spring and fall and as time permitted during the rest of the year.

The bitter past experiences with severe winters, the gradual overutilization of the forage resource, and the development of supplemental irrigation for meadowlands made hay making the focal point of livestock production in the Great Basin. Ranchers spent half of each year growing, harvesting, and storing hay, and the other half feeding it to their wintering brood cow herds. The generally accepted rule of thumb in the Intermountain area was one ton of hay to winter each brood cow.

The economic, social, and ecological changes that were the direct and indirect effects of making hay changed the lifestyle of the residents and grossly influenced the cold deserts of western North America. Hay production converted the range livestock industry of the sagebrush/grasslands from an extensive enterprise with minimum labor to a labor-intensive endeavor. Feeding hay increased labor requirements during the winter, but labor requirements peaked during the summer haying season.

The hay hands were largely a bachelor society, so the migratory aspects of their employment did not include the horrors of child labor, lack of educational opportunities, and self-perpetuation associated with twentieth-century migratory farm workers. The ranch bunkhouses with their full allotments of bachelors contributed to making Nevada the most "male" state in the union, with more than twice as many men as women, and the smallest proportionate number of women and children.[2]

The ranch hand created unique social and political conditions in the sparsely populated areas of the sagebrush/grasslands. A characteristic sight in Nevada from the 1890s to the 1940s was, in the words of social reformer Anne Martin, "the groups of roughly dressed men aimlessly wandering about the streets or standing on the street corners of Reno, Lovelock, Winnemucca, Battle Mountain, Elko, and Wells." They were in from the ranches with money to spend, and liquor, gambling, and women were their outlets. Each small rural town shamelessly flaunted a red-light district, usually surrounded by a high board fence erected as an illusionary barrier to protect the children of the townspeople and popularly known as the "stockade."

The migratory nature of the ranch hands and their almost complete lack of civic awareness removed a large portion of the potential electorate from the political process. This contributed to the relatively backward attitude of Nevada legislatures toward social reform. Nevada was one of the last western states to adopt the Nineteenth Amendment for women's suffrage. The very small voting electorate allowed various special-interest groups virtually to control the state.[3]

Social reformers such as Anne Martin bitterly attacked the injustices, real and imagined, of the seasonal employment of ranch hands. But ranch hands rarely thought of themselves as a discriminated-against minority. Ruggedly independent, they viewed themselves as masters of their own destiny, just as the ranch owners, whose market was controlled by relatively few meatpackers, treasured their self-image of independence.

The practice of "going down the road" provided a safety valve in rancher–ranch hand labor relations. The labor supply for ranching was usually insufficient, at least seasonally. If a ranch hand became dissatisfied on one ranch, he could always go down the road and find a job. Some employees spent a lifetime at a single ranch, while others regularly went down the road. If he could not find an incident to justify his leaving, the chronic quitter would invent a reason. Within a single valley or ranching area, ranch hands often made a circuit from ranch to ranch, eventually coming back to their original employers having forgotten the incident that sent them down the road. The circuit required anywhere from a few months to twenty years depending on the intensity of the incident that sent the ranch hand down the road and the memories of the employee and the employer. Through the ups and downs of ranch work, the men gloried in the fact that they did not have a union to tell them what to do.

More than one ranch hand went on a roaring drunk on the wrong side of the tracks in Elko, woke up on a haying crew forty miles from the nearest county road, and claimed he had been shanghaied. There are ranches in Nevada today that hire hands with the understanding that they will be returned to town after a month's work and allow no personal transportation at the ranch. If the ranch hand is going to go down the road before the month is up, he will do it on foot, and town is fifty miles away.

The ranch hand—part cowboy and seasonally a hay hand, ditch mucker, and fence builder—was a fixture of western agriculture from the 1890s to 1942. This was not a short-term phenomenon. The ranch hand society lasted for fifty years in the sagebrush/grasslands, where the grazing of domestic livestock is little more than a century old.

Any attempt at a broad-brush portrait of the ranch hand is doomed to failure. Excessive use of alcohol, limited education, skill in working with horses and his hands, but without the marketable skill of a carpenter or blacksmith all fit as generalities. A short spree of outrageous behavior that flaunted accepted morality provided both an identity and an outlet from weeks of poor living conditions and hard work. H. L. Davis, probably drawing on his own life as a drifter, described the feelings of a bachelor cowboy riding into an eastern Oregon town in the early evening, contemptuous of the families safe and snug behind lamp-lit windows and proud of the stares of disapproval that followed him down the street.[4]

Not all hay hands were bunkhouse bachelors. The seasonal nature of the work allowed smalltime miners and prospectors, woodcutters, homesteaders, and many Indians to work in the hay for a cash income to supplement their other activities. Haying was the first work experience for many boys, both from rural areas and from the towns. Almost every description of haying operations includes descriptions of activities performed by boys.

Some seasonal hay hands wintered in sunny California and rode the Central Pacific across the mountains to work in the hay. Some returned to the same ranch every year. Some were railroad tramps, although many tramps worked only under the most desperate situations, if ever. Henry Miller, who owned ranches from California through Nevada to eastern Oregon, developed a special policy toward such tramps. His rules were as follows:

1. Never refuse a tramp a meal, but never give him more than one.
2. Never refuse a tramp a night's lodging. Warn him not to use any matches and let him sleep in the barn, but only for one night.
3. Never make a tramp work for his meal. He is too weak before and too lazy afterward.
4. Never let tramps eat with the men. Make them wait and eat off dirty plates.

The Miller and Lux ranches became known among tramps as the "dirty plate route." Henry claimed that this policy saved him millions in fire insurance premiums.[5]

There was a definite class structure among hay hands. As a small boy,

J. A. Young learned this from water dippers. His grandmother's house had a well located on a side porch. The ranch hands used the well to fill their water bottles, usually gallon wine bottles covered with burlap sacks. On a porch post by the well hung a series of metal dippers. A ranch foreman explained to Young that the top dipper was for his grandmother and her house people. The next one down the pole was for the cook, foreman, and blacksmith. The next nail held a dipper for the seasonal hay hands. Lowest on the pole was the "happy" dipper, for the personal use of a mentally unbalanced hay hand named Happy for his perpetual smile. No one wanted to catch a case of the "happies" by drinking from his dipper.

The early Mormon settlers had a unique way of establishing hay rights on the common forage grounds. On the night before the hay was to be cut, a community party was held. At midnight the men adjourned to the haying grounds and began to cut a swath around the hay they proposed to cut. They could have all the hay they could surround by daybreak. If they set their sights too high and failed to close their piece by sunrise, anyone had the right to cut in their area.[6]

Hay making did not start suddenly after the hard winter of 1889–90. From the earliest days of ranching in the sagebrush/grasslands almost every ranch put up some hay for stock horses, but the quantity of hay produced relative to the number of cows wintered was such that only a small portion of the stock could be fed, and only for a short period. Some of the large ranches cut quite a lot of hay before 1889–90. According to the Elko paper, Pedro Altube of Independence Valley planned to cut eight thousand tons of hay during the 1887 haying season.[7]

A few early hay ranches were established on favorable sites along the California Trail, but local markets were necessary for significant agricultural development to take place in northern Nevada. These markets first appeared along the Humboldt mining district in 1860 and 1861. California-bound emigrants passing through the Great Basin recognized the natural hay lands adjacent to such streams as Goose Creek, Thousand Springs Creek, and the Truckee and Carson Rivers.[8]

Hay found a ready market with emigrants and teamsters. The vast stretches of desert without sufficient forage to support draft animals by overnight grazing and cold, snowy winters when little forage was available even in the irrigated valleys made the conservation of forage as hay a necessity. Demand for hay grew during the 1850s, as western Nevada attracted settlers, and then skyrocketed during the 1860s mining boom. Thousands of draft animals were required to move supplies to the mines and ore to the mills, and a hay production industry developed to feed them. Hay ranchers in western Nevada faced a crisis shortly after 1868 when the Central Pacific Railroad completed its line into Reno, reducing the number of draft animals needed to haul supplies over the Sierra Nevada to the Comstock Lode.

Hay is a bulky product that is difficult to compress. Stationary hay presses existed in the nineteenth century, but the degree of compression obtained was not great and the process was expensive. The ranchers of the Truckee Meadows area specialized in producing hay for use at remote mining sites. Alvaro Evans shipped baled hay by railcar on the Central Pacific and then by narrow-gauge railroads to Austin or Eureka, Nevada. The bales produced by a stationary hay press were large, with fourteen bales constituting a carload.[9]

The hay farmer generally needed a market relatively close to his hay fields. Obviously, not every hay field could be close to mines, transportation centers, or logging operations where large numbers of draft animals required forage. The solution was to take the animals to the hay— not draft animals but beef animals that could walk to the hay and then walk to the railroad for shipment to market.

Agriculture during the three decades following initial settlement in the territory that was to become Nevada was perennially subordinate to mining. The preeminence of mining after the discovery of the Comstock Lode and the subsequent extension of the mining frontier eastward throughout the state overshadowed the related growth of irrigated farming and livestock production. Production from the Comstock silver mines declined precipitously after 1876, but desperate Nevada residents found that a thriving range cattle industry had developed in excess of what was necessary to feed the mining population. According to reports published by the state surveyor general, hay production in Nevada climbed from 65,900 tons in 1873 to 618,000 tons in 1900. In 1873 Elko County had 15,000 acres under cultivation for hay production with an average production of 1 ton per acre. By 1880, 16,000 acres were devoted to hay production in the county and production had climbed to better than 2 tons per acre. The climb in production probably reflects the development of irrigation systems. After the hard winter of 1889–90 the hay acreage shot up to 239,000 acres in Elko County and production per acre dropped to about 1 ton per acre, a level of production that was maintained for the next half century.[10]

Truckee Meadows was a center of hay production for the Comstock Lode area; after the decline of mining, the meadows became a cattle-feeding area. The Central Pacific Railroad stimulated the winter feeding industry in the Truckee Meadows by allowing cattle firms to ship animals to Reno, feed them on hay during the winter, and then reship them to California at the rate charged for a simple Nevada-to-California transfer.[11] In 1870 Reno hay ranchers advertised their crops statewide in newspapers, offering to feed cattle at a per-animal rate or to sell hay by the ton. The herds were shipped by rail or driven to the host rancher's land. Hay production reached 5,000 tons, with only 1,000 tons finding a market in draft animals. Without winter feeding, prices were expected to fall from the average of twenty-five dollars per ton, but the development of feeding stabilized prices at twenty dollars per ton.[12]

Ranchers who sold hay to teamsters often sold their entire crop to one purchaser and were paid in installments as the feed was delivered during the following fall and winter. Many hay ranchers contracted as teamsters in the off-season. They had to have horses and wagons to put up hay, and contract hauling was a natural extension. Contracts for hay and freighting were made on the basis of regular delivery to freighting companies, stage lines, and other ranchers.[13]

Early hay makers were not particularly choosy about what they cut for hay. David Griffiths listed alkali bullrush, cattail tine, and spike rush among the important hay species. All of these species inhabit saline or freshwater marshes and seasonal lakes.[14] To this list of unusual hay species we could add an occasional willow shoot and rabbitbrush stem. Professor Frank Lamson-Scribner, commenting on western hay species in 1883, suggested that a brush scythe was a suitable implement for cutting many of the hay crops.[15] When questioned about the quality of his native hay, one old rancher replied, "When the snow is on the ground all they want is something to chew—the deeper the snow, the harder they chew."

Creeping wildrye was one of the most important native hay species. This diminutive relative of Great Basin wildrye is a much smaller grass with abundant scaly rhizomes, or underground stems. Early ranchers in Nevada knew the species as blue joint. Griffiths considered creeping wildrye the most important grass in the Intermountain region. He compared its manner of growth to that of western wheatgrass on the high plains, another rhizomatous species. Griffiths found magnificent stands of creeping wildrye along the Humboldt and Quinn Rivers. It was growing on rich, nonalkaline, heavy-textured soils, and when properly irrigated yielded two to two and one-half tons of hay per acre.[16] Creeping wildrye was the only native grass that Griffiths considered adapted to the wild flooding method of irrigation, which left the low spots in the field covered with relatively deep water for extended periods. Of the introduced grasses, only redtop was sufficiently naturalized to warrant comment by Griffiths, who despaired at its failure to dominate the hay meadows.

Willows are an integral part of hay meadows. Where stream gradients flatten in natural meadows, the streambeds tend to meander in s-shaped loops. In time, the streams cut through the loops, leaving oxbows that provide habitat for strips of willows. In late spring when the meadows were being flooded before hay harvest, the herbaceous vegetation was so green that it assaulted senses made hungry for contrast by endless sagebrush gray. The willows, arranged in sensuous curves, provided a welcome contrast to the pale green wiregrass.

The introduction of a single forage species, alfalfa, changed livestock production and the economy of the entire Intermountain area. Because alfalfa is a legume that supplies its own nitrogen through symbiotic fixation, and because it had to be deliberately irrigated rather than wildly flooded, alfalfa outyielded the wild hay by three to four times. Wild hay was usually cut only once a year, while alfalfa might be cut as many as three times.[17]

Probably native to central Asia, alfalfa has long been associated with developed agriculture. It was spread in the Western Hemisphere by Spanish colonists. Historians disagree on the exact date, but sometime between 1851 and 1854 a party of gold seekers on their way to California stopped in Chile after rounding Cape Horn and saw alfalfa being cultivated. No one knows whether this party realized that Chile and California enjoy similar climates or it was a matter of chance, but alfalfa seed was imported to California.[18]

George Stewart quotes E. J. Wickson, a former director of the California Agricultural Experiment Station, as indicating that W. E. Cameron had a field of alfalfa under cultivation near Marysville, California, in 1851. Stewart also indicates that Mormon emigrants on the way to Utah by way of California were responsible for introducing alfalfa to the Intermountain area. The innovative agriculture genius Henry Miller is credited with being the first individual to widely cultivate alfalfa.[19]

Myron Angel considers alfalfa production to have started in Nevada as early as 1863. J. P. Petigrew planted four acres of "Chili Clover" along the Humboldt River in 1864. By 1870 alfalfa had become the standard forage crop on upland irrigated areas. It is important to remember that alfalfa never adapted to lowland sites along streams and never replaced native grasses in meadows. In 1879, thirty-five thousand tons of alfalfa hay were cut from an estimated twenty thousand acres in the Truckee Meadows, a sevenfold increase over the 1870 yield.[20] Fred Dangberg, the astute Carson Valley rancher, was quick to grasp the value of alfalfa and is often credited with being the first to widely cultivate the forage crop in Nevada.[21]

John Sparks was instrumental in developing alfalfa fields for the Sparks-Harrell Company. In 1896 the H-D Ranch in Thousand Springs Valley and the Hubbard and Vineyard Ranches on Salmon Falls Creek had huge fields of alfalfa.[22] One major problem with the alfalfa introduced to northeastern Nevada from Chile was its poor winter hardiness. Wendelin Grimm, a native of the German Grand Duchy of Baden, introduced the first winter-hardy alfalfa to the United States.[23]

Years of hard labor and great expense were required to establish fields of alfalfa; it was a very different proposition from flooding native mead-owland. Grubbing out the sagebrush cost $2.00–5.00 per acre; plowing the new land $2.50–4.00 per acre, plus $2.50–3.00 per acre for disking, and an additional $0.50 per acre for harrowing.[24]

Alfalfa did not grow well on the lowland meadows where the native hay was produced. Silt-loam-textured alluvial soils on fans and terraces were ideal sites for alfalfa production if irrigation water could be provided. The water had to be applied evenly. Alfalfa would not tolerate being flooded four feet deep in low places with the high spots taking care of themselves. This meant that after the sagebrush was cleared and the land plowed, some attempt at leveling had to be made. The only source of power available for that was horses. Crude wooden box levels were dragged across the fields after Fresno scrapers were used to move soil for major fills. Many of the alluvial fan soils were at least slightly alkaline. If small knolls were left in the field, they quickly became alkali spots as the capillary rise of irrigation water deposited salts on the high spots while the rest of the field was leached.

Despite the cost, alfalfa production was highly profitable for many ranchers. In the Lovelock Valley of Nevada, alfalfa land sold for thirty-six dollars per acre in 1900 and could be paid for in five years. Yields of alfalfa hay were five to six tons per acre. Production costs were estimated at one dollar per ton. In 1899, forty thousand tons of alfalfa hay were fed to steers before they were shipped to the San Francisco market.[25] Raising high-quality alfalfa hay and using it to finish steers before they were marketed circumvented the environmental constraints of the sagebrush/grasslands and allowed ranchers to meet the changing demands of meat consumers. John Sparks could use registered Hereford bulls to upgrade his Longhorn cows, put two- to three-year-old steers in wet meadows to put on flesh, and then finish them on alfalfa hay before shipping them to California markets.

For seventy-five years economists pointed out that native meadows produced only one ton of hay per acre while alfalfa fields averaged at least three tons per acre of much higher quality hay. They uniformly considered native hay production a waste of valuable irrigation water. Social activists applauded such observations, believing that alfalfa production by small farmers could break the "big" ranchers' hold on the economy of the Great Basin.[26] There were and still are two problems with this view. First, the heavy sod of the native hay lands was extremely difficult to till for alfalfa. The undulating meadows required considerable cutting and filling to obtain even water distribution, and often drainage had to be provided. The second problem was what to do with the production of the meadow hay lands if they were broken into small farms. The climate of much of the sagebrush/grasslands limits agricultural production to forage crops that are of value only when marketed through livestock. Livestock production in this environment requires rangeland as well as farms. How to integrate extensive rangeland livestock production with intensive farming was and still is a critical problem.

Slightly over one hundred years ago, James MacDonald remarked to a western newspaper editor that cattle and grain production received about equal attention from agriculturists in Scotland. He was amused to find the next morning's issue of the editor's paper informing readers that one-half of Scotland was pasture and the other half was cultivated for crops.[27] It was impossible for even a literate newspaper editor to consider that the two could be integrated parts of a single agricultural system. It is a basic fact of life in northern Nevada that 50 percent of the forage base for livestock production is derived from rangelands and 50 percent from irrigated lands that constitute only 4–6 percent of the landscape. After one hundred years, the integration of peak agronomic efficiency with optimum range and animal science technology is a necessity that is no longer just amusing; it is essential.

Haying operations normally started after the Fourth of July and finished by August 15. In northern Nevada, there was always a rush to be finished by the opening of the Elko County Fair in late August. The fair awarded ribbons for canned fruit and produce from the gardens of ranchers' wives, and featured horseracing for the wealthier ranchers and more venturesome cowboys. The fair also provided the means and the excuse for a roaring drunk by the hay hands.

The entire haying operation was conducted with a sense of urgency. Irrigation water was turned out of the fields, and haying began as soon as they were more or less dry. If the sedges and wiregrass dried completely, they were impossible to cut with the mower. Many ranchers raised grain on irrigated fields located on the benches. This grain, oats, or barley had to be cut and threshed after the haying was completed.

The were five basic steps in the haying process: cutting a standing forage crop, drying it, gathering it, transporting it, and storing the resulting hay. All of this was done with men, horses, and simple mechanical implements. Resplendent in its ornate cast iron, the horse-drawn mower started the haying. The head mower opened up the field, cut the back swath, and then was followed by the lesser-skilled operators cutting endless swaths around the initial piece.

The first mower in Elko County was a wooden-framed Buckeye purchased for $225.00 by Matthew Glaser of Halleck, Nevada. At the time, the cost was considered prohibitive. When the cost of mowers eventually dropped to $110.00, virtually every ranch in Nevada had a supply of mowers. The junkyard at the Walti Hot Springs Ranch in central Nevada contains the remains of eighteen mowers, and this ranch had a relatively small acreage in hay. Glaser's Buckeye mower was considered so valuable that it was disassembled at the end of each haying season. The parts were oiled, painted, or greased, and then hung in the barn to await the next rush of haying.

When the reaper was first invented—independently by Cyrus McCormick and Obed Hussey—it was intended for cutting grass as well as grain; there was no distinction between reaper and mower. Gradually, however, two different types of machines were developed, and by 1854 there was a clear distinction between them. The cutting apparatus on a good reaper consists of a series of triangular section plates, or blades, sharp on the two exposed edges and fixed side by side like large saw teeth on the steel sickle bar. The sickle bar is given an oscillating motion as the mower moves forward, and the sickle plates shear the herbage stems against the plates in the fixed guards that are rigidly attached to the bar that supports the sickle. The forward motion of the mower wheels is transferred to the horizontal oscillation of the sickle by a system of gears, a heavy flywheel, and a pitmans rod. The heavy flywheel helps smooth the jerks of the traction power.[28]

There was nothing boring about mowing hay. The finger bar and the horses demanded the operator's constant attention, and hidden irrigation ditches gave the unsprung mowers tremendous jolts. The mower seat was a piece of sheet metal attached to a four-foot piece of spring-leaf steel. The resulting ride approached that of a bucking horse. The seat on the mower was positioned so the weight of the driver counterbalanced the weight of the tongue, relieving the horses of this extra burden. Mowing was dangerous, too. More than one operator made the mistake of standing in front of the sickle bar only to have the team run away. Mower teams had a special affinity for taking the mower, minus a few parts, "back to the barn door." Archie Bowman, general manager of the Utah Construction Company ranches, happened upon one of the company's veteran ranch hands picking up pieces of a brand-new mower in a hay field one day. The old cowboy had started work for Sparks and Tinnin in 1884. Trying to control his temper, Bowman asked the tough ranch hand, who had spent his working life fighting against the Industrial Revolution, how in the hell the runaway had started. The fact that his boss was responsible for fifty thousand cattle running on 3 percent of Nevada's total land area did not matter to this man. After a chew, spit, and distasteful squint, he offered, "I guess they started about dead even, Mr. Bowman."[29]

The tendency for the teams to run away required constant vigilance, but the mower operator's attention was diverted by hordes of mosquitoes that swarmed up from the vegetation as the mower passed. The hay fields were also infested with horse and deer flies that left painful bites and could lead to runaways when the insects excited the horses. In some fields, the horses had to be covered with sacks to protect them from biting insects. It was perhaps a good thing that haying camps and bunkhouses did not include facilities for bathing, because the mower operators needed all the natural repellant they could muster. The hay itself formed a natural distraction to the mower operator. On windy days, and especially if the hay was tall, the waving of the grass produced a hypnotic effect that led to either sleep or seasickness.

The drag on the cutter bar as the hay passed through the long, pointed guards to be cut by the oscillating sickle plates was hard on the team, and it was customary for the lead mower to stop every few rounds to allow the horses to rest. Long after mowers were mounted on tractors, it was not unusual to see some weather-beaten old-timer sitting on his mower with the engine idling while he stared off into space, resting his horses.

Some ranchers in Nevada teamed an unbroken horse with a gentle horse on the mowers. After a summer spent fighting the drag of the bar, the unbroken horse would be fit for any type of work. Many ranchers considered horses cheap and never babied the workhorses. A green team was hooked to a mower and away they went. Care was taken to not let the team run away, however, for once this happened they were often ruined. To prevent runaways, green teams were worked with a trip rope attached to their legs. If the horses started to run, the mower operator could pull the rope and the horses would fall. JIC bits were often used on new teams to help control them. These two-barred bits damaged the horses' mouths but helped prevent runaways.[30] The classic literary account of Intermountain hay making is by H. L. Davis, who matched a racehorse against a mustang in a mower race in the Malheur Bottoms of eastern Oregon.[31]

Rocks and sticks were the bane of mower operators. If a hard foreign object in the hay passed through the fingers of the guards, it was likely to jam the sickle bar or, worse, knock out a section plate. The mower operator had to dismount and risk a runaway to free the jammed sickle. In fields with a lot of rattlesnakes, it was dangerous to dismount and fumble around in the hay.

After supper, mower operators turned their attention to maintenance work. The sickles required careful sharpening on a grinding wheel. The wheel was pedal powered, and the face of each section plate had to be carefully ground. The shower of sparks accompanying the shriek of grinding steel section plates was a characteristic part of warm summer evenings during haying time. This was such an important job that large ranches had a full-time sickle sharpener.

Major repairs were done by the blacksmith, but the mower operator was responsible for oiling and adjusting the machine and changing the sickles. Every mower had a built-in toolbox that contained an oilcan; spare guards, section plates, and rivets; and a cast wrench. Mower manufacturers must have competed to see who could design the most ornate and complicated wrench. It would have one fitting for the nut that held the pitmans rod to the head of the sickle bar, another to tighten guard bolts, a third for wheel nuts, etc. In a pinch, the wrench could drive staples in the gatepost knocked down while driving into the field and club a rattlesnake stuck in the sickle bar. An 1870 owner's manual for a Meadow King mower indicated that that year's model was furnished with two knives (sickles), two extra guards, two extra sections with rivets, a screw-wrench, an oilcan, a neck yoke, and a doubletree.[32] The oilcans that came with these early mowers were often handcrafted from copper or brass and are now collectors' items.

To start on a one-hundred-acre field of native hay with a five-foot sickle bar was the height of impertinence. It took three-and-a-half hours for each mower to drop an acre of hay. If a rancher was going to winter cows, he needed at least one thousand acres of native hay. This works out to thirty-five hundred hours of mowing, or 350 ten-hour days. To complete haying in a month, thirteen mowers had to be in operation six days a week.[33]

After the mowed hay had wilted, it was gathered into windrows with a dump rake or sulky rake drawn by a two-horse team. The rake operator tripped a lever with his foot to raise the rake tongs and dump the hay in windrows. If the hay crop was very heavy, this required a strong leg. The dump rakers could always be identified in the bunkhouse because their legs twitched all night long. Hay could be bunched directly from the mower swath with a buck rake, but the sickle bar had to carry a windrow attachment for this to work.

Raking required less skill than mowing and often was the job of boys who had gained some experience driving derrick horses. The sulky rakes were pulled by lighter horses at a trot. The rough surfaces of the native hay meadow made this a rough and dangerous ride for the young operators. J. A. Young can vividly remember the ashen faces of the hired men when they returned from finding the body of the neighbor's son in the hay field. The rake team had bolted and he had evidently bounced off the sulky at the right time to go under the dropping rake tongs.

After cutting and raking, there were alternative methods for completing the hay making. The method followed had much to do with the quality of the product obtained. If high-quality alfalfa hay was being prepared, it was gathered into shocks.[34] The shocking was done by hand, and it was the largest labor requirement of the haying process. Crews of twenty or thirty men shocking hay were a common sight. This was a low-skill job that required only strong hands and arms to grip a pitchfork. The purpose of shocking was to reduce the amount of hay exposed to the sunlight while the drying process continued. Sunlight destroys the green color of the forage and also the carotene pigments that are precursors for vitamin A. Perhaps ranchers in the 1890s did not know about vitamin A, but they did know that cows did best when they ate hay with a good green color and that special aroma that goes with correctly cured hay.

When making good-quality alfalfa hay, it was important to shock as soon as possible after cutting. On well-drained land, it was not uncommon to see two mowers at work in a field, followed only a few swaths behind by a two-horse rake; meanwhile, several men were making shocks as rapidly as possible so that practically all the alfalfa cut on a given day was in the shock that night. Cut at one-tenth of full bloom and properly cured, alfalfa hay kept most of the leaves important in the protein content of the hay. Timing and the type of shock were important for the correct drying and preservation of leaves. Each shock should be the size that a pitcher could lift on the wagon with one forkful.

High-quality alfalfa was moved from the field to the stack yard by slips or wagons. When the stack yard could be reached without crossing roads and the hauling distance was short, slips could be used instead of wagons. Slips were homemade contrivances consisting of eight one-by-twelve-inch boards, sixteen feet long, nailed together with a crosspiece at each end. A doubletree at one end provided the means of attaching the team. Pitching hay onto the low slips required much less labor than tossing it onto high wagons. The slips simply slid over the hay stubble following the team to the stack. Sometimes small iron wheels were added with a pipe axle in the middle of the slip to make it easier to pull.

Hay wagons were a much more common method of transporting hay from the field to the stack. Hay wagons were sixteen to twenty feet long with a smooth bed three to four feet off the ground. At each rear corner there was a diagonally braced post four to six feet high to help hold the load. A high seat was mounted on the wagon front where it could be reached from the top of the load. Below the right side of the seat was a metal brake lever, and nailed to the back of the wagon seat was a forked stick or V-shaped board for tying the reins. This was the domain of the teamster.

Teamsters and pitchers worked together, usually with two pitchers throwing hay up to the teamsters, who arranged the load. Teamsters on hay wagons were a semiskilled lot, while pitchers only had to be, in the rancher vernacular, "hell for stout." This combination often resulted in a running warfare between the young, practical-joker pitchers and the older, salty, humorless teamsters. No matter where the pitcher threw up his shock it was wrong as far as the teamster was concerned. Teamsters took a very dim view of foreign objects such as rattlesnakes or skunks coming up with the hay shocks. A little humor helped to make the long, hot days of backbreaking labor bearable.

If low-quality grass hay was being prepared instead of alfalfa, a different procedure was followed. Buck rakes or sweep rakes required the least labor and were therefore the cheapest method of transporting hay to the stack. A buck rake consisted of several long wooden teeth lying almost flat on the ground, pointed at one end and fastened to a strong framework on the other. In Elko County, the teeth were often made from aspen trunks. A teamster drove the rake with its teeth extended down a windrow until hay was piled against the framework on top of the long teeth. Then he raised the teeth slightly off the ground with a ratcheted lever and drove the load to the stack yard. Ordinarily, the rakes were twelve feet wide. With four-horse teams, larger rakes were used. Buck rakes required highly trained horses able to back up as well as go forward and stop on command. After green horses had been taught rudimentary manners by pulling mowers, they were transferred to buck rakes and the mower driver started on a new team. The buck rake evolved in the hay fields of the Great Basin. Early versions lacked a seat, and the driver walked behind the team. Eventually a modified frame evolved with a single rotating wheel in the back and a seat for the teamster. Buck rake teamsters were skilled operators and commanded a better-than-average wage in the hayfields. An unskilled operator might drive the wooden teeth into the rough surface of the hay meadows or fail to pick up the hay.[35]

Hay handled with buck rakes became a tangled mass. The rolling action of the leading knocked the leaves from alfalfa hay. With careful attention to timing and a minimum of rolling of the hay during loading, high-quality hay could be produced with a buck rake, but this implement was characteristically used for transporting low-quality or wild hay in the sagebrush/grasslands. During the ten-hour day, one man with a wagon could haul the hay from two acres; with a slip, three acres; and with a buck rake, four acres.[36]

The low annual precipitation in most of the valleys of the Intermountain area, and snow instead of rain in the winter, made it possible to stack hay outdoors. This was fortunate in view of the scarcity of lumber for barns. Spoilage occurred even under semiarid conditions, so the basic problem was to stack the hay as high as possible and to thatch the top well to minimize the loss. The height of the haystack became a measure of technological advancement. Horses supplied the power, and various types of slides and derricks the mechanical lift.

Homemade derrick stackers came in a great variety of designs. The most numerous derricks were the "Mormon" type, characterized by the boom that pivots on the top of the mast. The second most popular type was the mast-and-boom derrick, on which the boom extends from the side of the mast. These two accounted for 90 percent of the derricks, with cable stackers and tripods accounting for the rest.

During lifting, the hay was held by a Jackson fork, a sling, or, more rarely, by harpoon forks. The Jackson fork consisted of a triangular hardwood frame with four long, curved metal fork tongs. The tongs, in an open position, were forced down into the hay load, and the frame was snapped shut with a metal catch where the dumping rope attached. Once the teamster on the wagon had the fork forced into the load and the catch snapped, he shouted, "Take-her-away." The derrick boy walked his horse a prescribed distance, pulling the cable through a system of pulleys on the derrick frame, which lifted the fork and hay. The derrick arm swung to the stack, where the stackers shouted, "Dump-her," and the teamster tripped the dump rope, opening the fork and dropping the hay. The derrick boy unhooked and returned the horse for another load. The teamster pulled the boom and fork back with the dump rope. The general shouting of orders with tobacco-filled cheeks and the short attention span of derrick boys caused more than one teamster loader to lose a finger in a Jackson fork catch due to an early start.

The dimensions of the nets used for lifting hay varied depending on how the hay was transported to the stack. If the hay came to the stack on wagons, usually two nets were used per wagon. These nets were approximately nine feet by ten feet, so two fit on an eighteen-foot wagon. The supports for the net were constructed of wooden poles or, occasionally, pipe. The net was made of three-quarter-inch to one-inch-diameter manila rope. The support poles on each end of the net had rings where the cable from the derrick attached. Down the center of the net were hooks that could be tripped by the trip rope to dump the load.

One method of rigging commonly used to lift hay off wagons was to attach one end of the derrick cable to the tip of the boom. The cable looped down to the wagon and back up to the boom, where it fed through a pulley to the derrick and then by a system of pulleys to the derrick horses. Within the loop that dropped down from the boom were two special pulleys with hooks and special flanges that prevented them from fouling each other. The hooked pulleys were attached to the two sides of the net by the rings; as the cable tightened, it pulled the net closed as it was lifted.

When hay was spotted at the stack with a buck rake, a larger net of the same basic design was used. The net was as wide as the buck rake. The net tender pulled the boom back from the stack by the trip rope, and the derrick boy backed the horses to lower the net down the side of the stack. When the net was near the ground, the derrick boy stopped and the net tender unhooked the center of the net, which was then lowered to the ground. The net tender pulled the outside half of the net away from the stack and hooked the ring over a pin driven into the ground that held the net in place while the buck of hay was spotted on it. When the buck rake backed off the net, the net tender flipped the lifting cable over the pile of hay and hooked the cable to the ring on the ground peg and the one on the free end of the net. He signaled the derrick boy to start the horses; as the cable tightened, it closed the net. As the net full of hay was lifted, the derrick boom automatically swung toward the stack. The net tender let the dumping rope trail through his hands until the stacker shouted, "Dump her." The net tender jerked the rope and the net opened. The dump rope then became the haul-back rope, and the net tender pulled the net, boom, and cable back to their starting place. It was important that the net tender reverse the momentum of the swinging boom and immediately begin pulling the boom and net back. More than one highly paid stacker was knocked off a high stack or killed outright when the boom and net swung out of control.

George Stewart considered the ideal hay crew for a single derrick to be two teams with drivers to build the loads, two pitchers who remained in the field, one man to work the Jackson fork, two men to build the stack, and one boy to drive the derrick horse. One wagon would stand at the stack being unloaded without a driver. Stewart calculated labor costs with this crew as 3.2 man-hours and 2.1 horse-hours per ton of hay stacked.[37]

Cable stackers were used when enormous stacks were desired. A steel-wire cable anchored securely at both ends and supported by two pairs of heavy poles supported a small trolley that carried the Jackson fork or sling over the stack. Because of their inherent strength, slings were often used with cable stackers. The rope-and-pole sling was spread on the wagon before the pitchers threw up the shocks. The entire load was lifted at one time at the stack. When slings were used on a derrick, the derrick had to be supported with guy wires. Besides the derricks, there were numerous "patented" stackers manufactured by equipment firms and generally named for the blacksmith who had patented the design. The swinging Jenkins hay stacker and the overshot stackers by Dain or Jackson are typical of these designs. They required a capital investment of two or three hundred dollars and, being of eastern origin, were often not strong enough for western haying. In contrast, a blacksmith in Ontario, Oregon, made the necessary ironwork for a derrick for twenty-five dollars, and the rancher could supply and assemble the woodwork himself. By establishing a reputation for quality, this blacksmith developed a market for ten to fifteen derricks a year.

A modern traveler in the sagebrush/grasslands can spot an old hay field in the distance by the lonesome mast of the long-abandoned hay derrick. It is likely to be the first thing visible, long before the field itself is apparent. Sticking fifty to seventy feet into the air, such poles provide highly visible landmarks and beg an obvious question. In this environment where a cowboy has to ride a hundred miles to find a tree sufficiently large to provide shade, where did the ranchers get them?

When Utah Construction operated the ranches that John Sparks founded in northeastern Nevada, the company employed a full-time pole cutter named "Sanitary Bill." His name derived from a lifetime spent without bathing. Bill spent the summers in the Sawtooth Mountains of south-central Idaho cutting lodgepole pines for the company. He had a list of specific parts needed for derricks, gates, buck rakes, and horse-breaking corrals. Equipped with an ax, broadax, saw, and auger, he spent the summer shaping parts and the winter touring ranches assembling the various items ordered.[38]

In the wild-hay meadows of the Intermountain area, hay was often stacked with slides—huge, strongly built inclined planes. The hay was brought to the bottom of the slide by a four-horse buck rake and deposited on a chain net. A cable was fastened to the net and stretched over the top of the plane and the entire stack. The other end of the cable was attached to the foretruck of a wagon hitched to a four-horse team. When the load had been pulled up and dumped at the proper place on the stack, the net was drawn back to the base of the plane by a single horse, adjusted, and reloaded. The four-horse buck load averaged about one ton of native hay, and a load could be run up on the stack every six to eight minutes when everything was in good working order.[39]

The popular beaverslide was a more complex type of slide. A load of hay gathered from the field with a buck rake would be placed on the hay forks at the base of the slide. The forks, which resembled the fork on the front of the buck rake, would carry the hay up to the top of the beaverslide, where the load would fall over the end of the slide onto the stack. A team of horses hitched to the cable used a pulley system to draw the loaded fork up the slide. When the stack built up to the top of the slide, it was "topped" by the stacker and the beaverslide was then moved forward about fifteen feet and a new bent was started. Some very large beaverslides built in Elko County were capable of lifting four bucks of hay at one time and making stacks fifty feet tall.[40]

A decided disadvantage of the slide methods of stacking, especially the simple incline, was the difficulty in making the resultant stack waterproof. With a ton of hay dropping every eight minutes, the stacker had little time to rearrange the hay to prevent holes that let in moisture. The height of the stack was limited to the height of the slides, generally 10 to a maximum of 15 feet, creating a long, low, loose stack of hay subject to heavy spoilage. In contrast, Griffiths describes alfalfa haystacks built with a derrick near Ontario, Oregon, that were 375 feet long, 28 feet across, and 75 feet tall.[41]

Local ranchers evolved their own particular systems of handling hay. On Willow Creek in northeastern California, one ranch used four mowers as a cutting crew. After the hay wilted, two sulky or tumble rakes bunched it into windrows. From the windrows the hay was shocked for further curing. The shocks were picked up by a two-horse buck rake and spotted at strategic locations in the fields called yards. After all the hay had been yarded, it was loaded onto wagons using a low beaverslide, four bucks to a wagon. The wagons were unloaded with nets. A team of horses was required to pull the derrick cable with these big nets. The team was attached to the derrick cart by a type of doubletree called a stretcher. The cable was attached to a cart or the running gear of an old mower, and the derrick boy sat on the seat to drive the team back to lower the nets to the wagon.[42]

At three dollars a day, the hay stacker or mower (pronounced *mou-er*) was often the most highly paid man on the crew. The mower placed, or "mowed," the hay on the stack. Unless each fork or sling of hay that reached the stack was carefully spread and interlocked, water spoilage would ruin the hay. There was always some loss on the top of the stack due to storms and on the entire outside surface of the stack due to bleaching. The larger the stack, the smaller the proportion of hay lost to bleaching. Proper stacking could also limit penetration of moisture from storms.

Bill Majors was the most skilled stacker in Big Valley, Lassen County, California. With a helper he could do the work of five men on the stack. He did not stay on the ranches where he stacked hay, but walked miles across the valley to and from his home on the Lookout Indian Colony each day. When breakfast was served at 6 A.M., he was there at the table.

Good stackers would often taunt the net or fork tender with calls of "more hay, more hay." If the net tender got overanxious and did not watch what he was doing, the trip rope might catch on the wagon or buck rake and dump the load on the way up, before it reached the stack.

Initially, the middle of the stack was the most important. The first several loads of hay were dropped along the middle line of the stack, with the flakes of hay flattened down and thoroughly trampled. The middle depth was three to four feet higher with a uniform slope toward the edges. After the stack was started, the outside of the stack became the most important part. Trampling was done so the hay was progressively looser from the middle toward the edges. Once the hay was in the stack, a crew was kept busy rebuilding the stack yard fences. These had to be completed before cattle were turned in to graze on the crop aftermath.

Their dependence on hay for forage to winter their cattle challenged the managerial abilities of ranchers, who had to raise an irrigated hay crop, direct a small army of hay hands, and obtain the necessary capital to finance such operations. There were many opportunities to increase both efficiency and the quality of the product produced. In the 1870s Jasper Harrell turned cattle loose on the pristine sagebrush/grasslands and marketed his increase with a minimum of interference with nature. By 1900 John Sparks was faced with the complexities of irrigated agriculture and making hay. Some of the Indians on Sparks's hay crews had probably once lived in the sagebrush/grasslands environment as hunter/gatherers. Time in the exploitation of environments is a fickle thing, with technology accelerating its passage. Haying in the sagebrush/grasslands was born out of necessity and persisted as a labor-intensive, horse-powered enterprise for fifty years.

CHAPTER 10

From Dugouts to Cattle Empires

When Texas ranchers first moved up the plains to the Northwest, ranch headquarters often consisted of a sidehill dugout. But as the cattle industry flourished, ranch headquarters evolved into an elaborate collection of buildings. In the 1890s, a major ranch headquarters had to be self-contained because ranchers were isolated for several weeks each winter and early spring. The H-D Ranch of Sparks-Harrell on Thousand Springs Creek received a shipment of supplies each fall before being isolated by winter storms, and again before haying started.

Ranch buildings were built for function, not beauty or comfort. The ranch headquarters usually included a blacksmith shop, leather and harness shop, dairy, smokehouse, root cellar, laundry, bunkhouse, icehouse, cookhouse, cook's cabin, chicken house, horse barn, milk cow barns, buggy house, sheds, corrals, and a big cast-iron dinner bell.[1]

The San Jacinto Ranch headquarters, which Sparks named for the battle in which Texas troops defeated Mexican forces and avenged the loss at the Alamo, exceeded normal standards for self-containment; it had a store. The white stone store with its tall windows and huge iron door and shutters was a controlling force in northeastern Nevada.

Sparks-Harrell and later owners sold supplies selectively. They influenced the development of mining and small ranches in northeastern Nevada by refusing to sell supplies to newcomers, forcing them to go to Wells, Elko, or Idaho for supplies.[2]

In all her years as a resident of San Jacinto, Nora Linjer Bowman never saw the inside of the bachelors' bunkhouse. It was not considered a fit sight for the general manager's wife. The structure dated to Sparks-Harrell days, and her description would have been interesting. If you have seen one, however, you have a pretty good idea of them all—potbellied woodstove for heat, wooden bunks, bedrolls with canvas covers on the bunks, a wooden table for card playing with benches or nail-keg chairs. Worn-out boots, stray gloves, winter sheepskin coats, spring gumboots, and half-finished braiding projects accumulated under bunks and in corners. Washing was done in tin basins at the cookhouse. A trail behind the bunkhouse led to an outhouse. On Saturday night when there was a dance at the schoolhouse, a quick talcum-powder bath and a newer shirt sufficed.

Every rancher had to enforce some control over alcohol in the bunkhouse. Often this took a passive form. Once a ranch hand had dried out from a spree in town, he was too broke and too far from town to get any more alcohol. Others were craftier and kept rations of sweet wine hidden in the outhouse or elsewhere around headquarters.

Bunkhouses were cold in the winter and ovens in the summer. Often they were dirty and infested with vermin. Many seasonal hay hands preferred to camp in "jungle towns" located near the ranch headquarters instead of sleeping there. The bunkhouse at San Jacinto was constructed of lodgepole pine and covered, originally, with a dirt roof. Dirt roofs leaked with every rain and continually dribbled fine particles of soil. To counter this, canvas was tacked on the ceiling. The false canvas ceiling provided entertainment for the crew, who enjoyed watching the movements of packrats as they slid down and scratched along the upper surface of the sagging canvas.[3]

The ranch blacksmith shop, which many associate with glowing forge and ringing anvil, contained the smells and artifacts associated with horses. On most ranches the blacksmith was also the horseshoer. Hooked over a pole fence or piled in a corner were hundreds of used horseshoes ranging in size from small, narrow shoes for pack mules to massive, wide shoes for workhorses. There were always oddball shoes forged to correct some injury to a horse's hoof. In the 1890s a sentimental ranch hand might have an ox shoe recovered from the California Trail nailed to the wall in honor of that time, a half century before, when oxen challenged workhorses as a source of power.

Ranch dogs, always abundant, loved to search the horseshoeing area for pieces of horse hoof—choice morsels. After the noon meal at the cookhouse, ranch hands gathered at the blacksmith shop to ruminate before returning to work. Dogs fighting for the best horse hoof chip enlivened these breaks.

Horseshoeing equipment included pincers in assorted sizes, cutting nippers, various weight-driving hammers, and a punch to make the nail holes in the shoes.[4] Roundups and haying were busy times for the blacksmith. Rank young horses had to be roped and thrown before their shoes could be applied. Handling horses in this way was rough and dangerous work that required skill. A good horse could be ruined or the ranch hand or blacksmith injured if the job was not done correctly.

Even broken horses could be dangerous to shoe. Some horses developed devilish schemes to make the blacksmith's life miserable. To examine a hoof, the farrier lifted the hoof and held it between his legs. By gradually increasing the weight placed on the leg the shoer was holding, the horse could bring a backbreaking load to bear without the farrier noticing. Many horseshoers had a mean outlook on life that probably developed from working all day bent over, supporting part of a horse while being kicked, bitten, stepped on, and generally roughed up. Many carried large rasps in their tool kits to teach manners to misbehaving horses.

If the horse was wearing shoes when brought into the blacksmith shop, the farrier examined the old shoes and the hoofs for signs of uneven or unusual wear. The old shoes were pulled with dull nippers and the hoof was cleaned with a horseshoer's knife. The hoof was trimmed and shaped with a six-row rasp. The new shoe was heated black-hot, not red-hot, in the forge and applied to the horse's hoof. The shoer noted the burns on the hoof and used his rasp to reshape the hoof until it was perfectly flat.

The shoe was then nailed to the hoof with big-headed horseshoe nails. The points of the nails emerged through the side of the hoof, and the farrier had to bend them over to secure the shoe. If the horse made any violent movements at this stage of shoeing, the shoer stood a good chance of having the nails stuck in his legs, although he was wearing a heavy leather apron to protect them. The bent-over nails were clinched with a sharp hammer rap while a clinch bar was held tightly against the nail-head. Sharp nippers were then used to cut off the nails, and a curled-face clincher bent the clinched nails firmly into the hoof.

In the fall, before the winter hay feeding started, the teams that were to be kept at the ranch headquarters to pull the feed wagons were shod with caulked shoes to give them traction in the winter's mud and snow. The blacksmith welded the caulks onto smooth shoes. Cans of Cherry Heat Welding Compound served as a flux when shoes and caulks were heated red-hot and welded together.

Once or twice a year, usually before haying, wagon wheels had to be reconditioned. This was the blacksmith's job as well. The wheels were removed and a rod was placed through the axle hole of each. The wheel, suspended by the rod, was placed in a sheet metal pan of boiling linseed oil and was slowly rotated until all the wooden parts were reconditioned with absorbed oil. If the metal rim was worn or loose, the blacksmith replaced it.

The blacksmith who worked for John Sparks had to be especially skilled because Sparks was a hot-rod buggy driver who reached for the whip when other drivers reached for the brake. In 1896 he was driving on the fourth set of wheels on his custom-made buggy when he gave banker Jackson Graves a hair-raising ride from the H-D to the Vineyard Ranch.[5]

The hubs of all wheels, from buggies to hay wagons, were lubricated with axle grease. This hard, sticky grease came in a gallon pail that was saved after it was empty for use in measuring. "About an axle grease bucket's worth" was common ranch English.

The blacksmith shop was the center for harness and saddle repair. A workhorse's harness was largely leather, but it required a bewildering variety of snaps, buckles, and chains to function. Harness supplies included utility cockeyes, harness snaps, dixie belt snaps, breast strap roller belt snaps, open eye snaps, and snaps and thimbles. For harness repair, a ranch hand sat on a wooden workhorse that supported a long-jawed wooden vise at one end. Heavy needles, awls, waxed thread, leather punches, and knives were his tools.

The harness was attached to the horse collar with a pair of curved hardwood sticks called haines. Commercial teamsters often decorated the tops of their haines with curved wooden bows hung with bells. On ranches, the haines usually ended in polished brass spheres. Just as truck drivers today polish the chrome on their rigs, teamsters took pride in polishing haines ornaments. The teamster sometimes added ornaments to the headstalls of his team's bridles even though the harness and horses belonged to the ranch. If the teamster wintered in California, the ornaments were conchos made from silver dollars or Mexican pesos by Mexican silversmiths. A more common practice was to salvage the back of a broken pocket watch and attach a wire loop to the back by filling it with lead or solder. Simple things, these ornaments, but they were symbols of pride.

By the 1890s most of the cowboys in northern Nevada and southern Idaho rode single-cinch saddles with Walker trees made by the Visalia Saddle Company of San Francisco. The leather used in these saddles was oak-tanned and at least seven years old when the saddles were made. Saddles and horse equipment evolved considerably during the three decades of intensive livestock production on the sagebrush/grasslands. Texans brought their equipment and traditions when they delivered the first cattle, and the Spanish vaqueros from Old California contributed a different style of equipment.[6]

Somewhere in the saddle room or blacksmith shop there was a chest devoted to horse medicine. Horses are prone to injuries, especially around the feet and legs. Bottles of Bone Blister, White Liniment, Badger Balm, Lucky Four Blister, Germ Killer, and Antiseptic-Poultice were part of the mystique of treating unsound horses.

The jewel of every large ranch headquarters was its garden. Although many ranch headquarters were at elevations between five thousand and six thousand feet, the gardens grew a broad range of produce. During the growing season the garden was the only source of fresh vegetables. It also supplied the preserved vegetables that were eaten during the rest of the year. Potatoes, cabbage, and root crops, especially carrots and parsnips, were stored in underground cellars for the winter. Many crops were canned in glass jars for use all winter, with botulism a constant risk.

In the spring, the half-rotten vegetables from the cellar were a poor excuse for food, but better than nothing. The mail-stage driver in central Nevada hated the springtime. In addition to the muddy roads, he had to face lunch invitations from ranchers—sometimes boiled, thin venison and rotten cabbage dug from the root cellar. The visit from the mail stage was important, and ranchers—and their wives—were eager to entertain the driver. Besides the mail and papers, the mailman delivered news that was too trivial and too personal for the local papers.[7]

The chore man tended the garden as well as the milk cows, chickens, and pigs. There were many jokes about Texas cowboys who had never tasted cow's milk. On late-nineteenth-century Nevada ranches the milk cows provided milk, butter, and cottage cheese for the cookhouse, and skimmed milk for the pigs. The late-nineteenth-century cowboy ate more cured pork than fresh beef. The pigs were slopped with table scraps and skimmed milk. Many ranches raised grain in rotation with alfalfa, and this grain was fed to the pigs. The pigs' rations would have been deficient in quality protein without the skimmed milk.

As far as the ranch hands were concerned, the cookhouse was the focal point of ranch headquarters. The cookhouse was located outside the kitchen and was a simple room with long wooden tables covered with oilcloth. The kitchen held massive implements large enough to feed forty men three meals a day, a huge black stove with tiered warming ovens, and a tank water heater. For breakfast, great black griddles placed on top of the stove produced hotcakes eight inches in diameter and one-half inch thick. It was a good thing there were few deep bodies of water in the sagebrush/grasslands; after a breakfast of hotcakes like these, swimming would have been difficult.

No matter how hot the weather, the cook still had to feed wood to the stove. Even if the men ate a cold meal, the stove had to heat water for washing the dishes. On small ranches, the herculean task of cooking for haying crews fell to the rancher's wife. One prominent Nevada ranch lady related a terrible nightmare in which she dreamed she was in hell cooking in midsummer on a woodstove for a forty-man haying crew. She knew she was in hell because when the crew came to sit down, each wore the uniform of one of the prominent federal land management agencies.

Besides giant hotcakes breakfast consisted of home-cured ham or bacon; mush, either creamed or fried; and dozens of fried eggs. At some ranches it was traditional to have beefsteak for breakfast. If the cook failed to have steaks ready for the crew, they would probably go down the road. Just as important as steaks at some ranches were the baking powder biscuits. The food was served on heavy ironstone plates along with gallons of boiled coffee served in handleless mugs so thick that the scalding coffee never cooled. Those who did not care for scalded tongues and tonsils slurped the coffee from a saucer. The old-timers on the ranch crew treated the cook with cold disdain. It was beneath their dignity to ask for a second cup of coffee; they simply tapped the empty mug with a knife handle.

The noon meal was simple but tremendous in bulk: boiled potatoes, fried thinly cut steaks, white or brown gravy, beans and ham hocks, and home-baked bread or biscuits with farm-churned butter. During the summer, lettuce was served wilted with vinegar and bacon grease dressing. A few fresh vegetables went a long way. Dessert was rice or canned peaches.

If the cook was feeding a large haying crew, he might serve beef every day. It took a large crew to consume a beef before it spoiled. The steaks had to be cut thin before frying. The grass-fed animals had little marbled fat, and the meat was tough. The teeth of older ranch hands were not up to chewing thick hunks of such meat.

Facilities for slaughtering beef were primitive at most ranches. Usually a windlass was arranged in the corral between two stout gateposts. The animal selected for slaughter was herded into the corral, shot or knocked in the head with a sledge, then hoisted with the windlass and bled. The offal was dumped on the ground and left for the pigs. In the early days of ranching, during the 1870s, the offal often went to Indian families that had become attached to the ranches when cattle outcompeted them for grass seeds. The hide was thrown over the fence, hair side down, to dry for conversion into rawhide. The meat was cut up and hung in a screen cooler. Small ranches often had cooperative exchanges of meat so it could be used before it spoiled. In the summer, the only alternative was to put the excess meat in brine as corned beef.

Beans were a standard item on cookhouse menus, and cooks who left rocks in the beans were not popular with the crew. The lazy cook's method of cooking beans was to dump them in the pot without cleaning them and let the chaff float to the surface and the rocks settle to the bottom.

The evening meal often featured a large roast or two main meat dishes like baked ham and roast beef and more boiled potatoes and gravy. The heat in the kitchen during the evening meal and through the cleanup afterward reached boiler-room intensity. During the late summer, door and window screens were black with flies waiting for an opportunity to pop in.

After the wreckage of the evening meal was disposed of, it was time for the cook to do the next day's baking. Besides bread, the crew thrived on huge slices of fruit pie—dried apricot and raisin during the winter, and apple in season. If the cook was in a good mood, he might prepare donut dough to fry for the next morning's breakfast. The kitchen started to cool late in the evening, especially if the ranch was situated at the mouth of a canyon where breezes rustled the cottonwood leaves. The cook had a few moments to sit on the porch and enjoy the evening before dropping exhausted on a bunk in a hot, stuffy room with mud dauber nests on the peeling wallpaper. At 4 A.M. he was up to light the barely cooled woodstove to start another day. Who were these supermen who ruled the cookhouse? Often the cook was a dried-up ninety-pound Chinese displaced from some Central Pacific construction crew by way of a worn-out mining claim.

The cookhouse was a significant part of the ranch's production budget. The cost of boarding labor for growing hay was $0.29 per ton, and the cost of boarding the haying crew was $0.63 per ton. This was in addition to the total labor cost of $3.68 per ton of grass hay produced.[8] The cash expenditures for supplies for the cookhouse included flour at $3.50–4.50 per hundredweight, sugar at $6.00–8.50 per hundredweight, beans at $7.00–9.00 per hundredweight, and coffee beans at $0.33–0.50 per pound.

Henry Miller, the master of nineteenth-century ranch management, looked closely at the vegetable garden when he visited a ranch. He chastised one of his managers with, "Your vegetable garden is not big enough for the men you have. Good vegetables make the men more content and draw a better class of laborers." He surprised another of his managers by examining the garbage bucket in the back of the cookhouse. "Those potato peelings are too thick. You can tell a good housekeeper by look-ing at the potato peelings."[9] The ranch manager had to balance the wel-fare of the crew with the cost of production. If he had the cook skimp on food, the crew went down the road.

Cooks were notoriously touchy. One cook posted a sign in the cookhouse: "If you can't wash dishes, don't eat. We use wood in the cookstove cut 16 inches long but no longer. A busy cook loves a full woodbox. A full water bucket makes a happy cook, stray men are not ex-empt from helping wash dishes—bring wood or water. The well is just 110 steps from the kitchen, mostly downhill both ways."[10]

Settlers' first reaction to the treeless, semiarid environment of the sagebrush/grasslands was to plant trees to modify the environment into something that fit their conception of desirable surroundings. Very few trees are adapted to grow in the sagebrush/grasslands. Much of the course of the Humboldt River from Elko to the Humboldt Sink was de-void of trees under pristine conditions. Scattered groves of quaking as-pens grow high on the sides of the fault-block mountains, but aspens did not grow well in the environs of most ranch headquarters, and quaking aspens are not easy to propagate. Other members of the poplar family did adapt to conditions around ranch headquarters and were widely planted. In the western Great Basin, Fremont cottonwoods form a natural gallery forest on the banks of the Truckee, Carson, and Walker Rivers and ex-tend well out into the Carson Desert. Black cottonwoods are native to streamside environments in the mountains of northeastern Nevada. Settlers found it relatively easy to propagate the native cottonwoods by taking root cuttings or sucker sprouts with roots. The rapidly growing cottonwoods provided welcome shade for buildings, cabins, corrals, and bunkhouses. The imported Lombardy poplar was widely planted around ranches and homesteads, but this relatively short-lived tree often suc-cumbed to drought and disease. The naked skeletons of Lombardy pop-lars mark innumerable abandoned ranches and homesteads.

The ranch headquarters was the only part of the sagebrush/grasslands ranching system where women regularly played a role. There were occasional female ranchers, and even outlaws, but men dominated ranching. The life of the rancher's wife was hard and lonesome even if she lived on one of the larger ranches. The distance between ranch headquarters on the sagebrush/grasslands precluded close contact with neighbors. Often the only companions available for the ranchers' wives were the Indian women who washed clothes and worked around the ranch house. Social attitudes and the language barrier usually prevented real friendships from forming.

A folk story told in northern Nevada describes the reaction of a Crowley Creek rancher's wife to the sagebrush environment. The Crowley Creek woman was left completely alone at her stone-walled ranch house for considerable periods while her husband supervised roundups. One year she begged her husband to have a party before he left for the fall roundup. The invited guests traveled a hundred miles across the sagebrush valleys and barren playas to attend. When all the guests had assembled in the ranch yard and pitched their traveling tents beside their rigs, the Crowley Creek woman went into the house, stuck a rifle barrel into her mouth, and killed herself. She left behind a note saying that she did not want to die alone in this empty land.

The winter haying crew was a permanent part of ranch headquarters. In contrast to huge summer crews that were strictly seasonal, the winter hay feeders were there year-round. After the summer hay crew departed to a warmer climate, the permanent crew often ate their meals with the foreman, or on smaller ranches with the owner. Hay feeding did not start until early winter. Cattle were gathered from the summer range in October and grazed on crop aftermath—the regrowth on the native hay meadows.

On Sparks-Harrell ranches, the cattle were worked and sorted as they were gathered. Dry cows and steers that were not ready for market were pushed out onto the salt deserts of the Bonneville Salt Flats northeast of Montello or, on the northern ranches, sent to the Owyhee Desert. Cows with small calves and replacement heifers were kept on hay aftermath, and when that was exhausted were fed hay.

If the winter was severe, weaker animals outside the fenced fields were worked into the fields and fed hay. In central Nevada during a hard winter, the cows and calves were kept in the feed yard area while the yearlings and two-year-old steers walked the fence trying to get in. The feeding crew on one ranch warned the foreman that the steers were getting weak and should be let in for supplemental feeding. The ranch foreman finally relented and sent the men to bring in the steers, but very few were found alive. One of the cowboys explained that the steers wore such a deep trail around the fence that the sides caved in and the steers disappeared.[11]

Under average conditions, with conveniently located stockyards and feed grounds, one man could feed and take care of four hundred head of stock cattle during the hardest winter months. Although one ton per brood cow was a general hay requirement, ranches that bordered salt desert areas might get by with one-half to three-quarters of a ton of hay per cow. In high mountain valleys such as the headwaters of the Humboldt River, Salmon Falls Creek, and Goose Creek, each cow needed one and one-half tons of hay.[12]

Some ranchers treated hay in the stack like money in the bank; they fed only as a last resort, and then as little as possible. Their cattle went on the range in poor condition each spring, but they had hay reserves carried over from winter to winter that were invaluable during severe winters. Ranchers who subscribed to this philosophy often fed most heavily during the early part of the winter, so the stock was in good shape for the most severe weather, and then greatly reduced the amount of hay fed during late winter and early spring, forcing the cattle to rustle for their own forage.

John Sparks was exceptionally miserly with his hay supplies. When Jackson Graves toured the ranches in 1896, the stockyards on the Salmon Falls ranches contained stacks that were three years old.[13] Only six years had passed since the disaster of 1889–90, and the memory of thousands of dying cattle was slow to dissipate.

When hay feeding started after Thanksgiving, it was not such a bad job. Once the cows got their hay and the horses were taken care of, the winter crew worked on other chores. On stormy days the crew sat around the stove in the blacksmith shop and repaired harness and saddles while they watched the ranch dogs steam before they grudgingly moved back from the stove. By March, the novelty was gone. Seven days a week it was the same thing every morning, snow, rain, or shine.

The tempo of ranch life quickened in the early spring. It was time to muck ditches, get the irrigation dam installed, and brush the cow chips on the meadows. Instead of being the focal point of the day, hay feeding was an extra chore tacked on top of everything else. The rancher was more than willing to turn the cows loose. For their part, the cows were tired of a steady diet of dry hay. The lure of fresh grass in the hills drew them away from the fields.

Twentieth-century range science has proven that grazing the same plant species at the improper season with the same class of livestock will result in the disappearance of the forage resource. Grasses native to the sagebrush/grasslands grow very little during winter due to the low temperatures. When it warms up in the spring, these grasses must use their carbohydrate reserves to renew growth while soil moisture is available. If native perennial grasses are grazed by the same class of livestock at the same time each spring, the grasses die. The earlier in the spring the grasses are grazed, the greater the mortality rate. Ranchers who fed heavy early in the winter and tapered off later were forcing cattle to use the perennial grasses before they had a chance to rebuild their carbohydrate reserves.

Ironically, feeding hay contributed to the overutilization of rangelands located close to the ranch headquarters. Lewis "Broadhorns" Bradley, a nineteenth-century governor of Nevada, once observed, "There is a good deal in educating a critter. He is like a man—if he knows his living will depend on his rustling, he'll rustle!"[14] The winter of 1889–90 proved that hay feeding was necessary for the animals' survival in some winters. However, feeding hay kept the stock concentrated near the feed grounds, which resulted in severe overutilization of native perennial grasses at the worst possible season. The concentration of human activities and grazing animals around the feed yards made the ranch headquarters the most degraded portion of the environment of the sagebrush/grasslands.

CHAPTER 11

Herefords in the Sagebrush

The bulk of American cattle during the mid-nineteenth century, especially those that belonged to the farmers of the frontier districts, were poor-quality animals. The Longhorns of Texas were virtually a free-breeding population; herdsmen made no conscious effort to improve their quality. The milk cows that followed the covered wagons to Oregon or California were generally plain-grade animals of poor quality as well. By the early nineteenth century, the concept of improved livestock breeds was in full blossom in England and Scotland. Just as the roots of the Texas Longhorn reach back to the open ranges of the Extremadura in Spain, the English cattle breeds that began to appear in the sagebrush/grasslands in the 1870s and 1880s extend back to the United Kingdom.

In colonial America, country gentlemen of the largely agricultural society followed the example of English and northern European landowners and made their estates showplaces of agricultural technology. An imported English bull or Belgian stallion provided status. This concept was transferred to the frontier through the local fair and livestock show. After the Civil War, the railroad transportation system had evolved sufficiently in the upper Midwest and the eastern portion of the country to

allow statewide and regional livestock shows. The increased public ex-
posure heightened interest in improved breeds of cattle, and the livestock industry responded quickly.

Nearly every rancher could see that the Longhorn had many draw-
backs as a red-meat producer. There was no practical way to replace the
Longhorns that had been stocked on the rangelands, but herds could be
gradually improved by removing Longhorn bulls and replacing them
with high-quality English-breed sires. This concept generated a huge
market for bulls.

Improved breeds did not win instant or widespread acceptance in the
West. When the Scottish agriculturist James MacDonald visited Texas in
the 1870s, he doubted that Shorthorns would ever be a success there
because few of those that had been introduced had survived more than
twelve months.[1]

The fertile Tees Valley in northern England was the cradle of the
Shorthorn. Here, late in the eighteenth century, was born a breed of
cattle destined to become the most numerous improved strain during the
nineteenth century and one of the most valuable of the world's breeds.[2]
The counties of York and Durham in the Tees Valley had long been famous
for a type of cattle known as the Teeswater breed. Teeswater cattle were
characterized by size of frame, strength of bone, and ability to attain great
weight at maturity. In color, these ancestors of the Shorthorn were of
various combinations of light yellowish red and white.

Throughout most of the eighteenth century, breeders of Teeswater
stock were generally in accord as to the type of cattle best suited to their
operations. It was not until the last two decades of the century, however,
that the breed began to take definite form. Molding of the Shorthorn was
initiated by the skillful hands of two tenant farmers in Durham, the
brothers Charles and Robert Colling. Their foundation material, secured
in 1783, comprised the bull Hubback and a number of cows purchased
at the Darlington market.

Charles Colling was a student of Robert Bakewell, the first man known to have bred livestock scientifically. Quick to discern the potential value of Teeswater cattle, Charles and his brother set out to improve and establish the type. The first ten years were spent experimenting with breeding material approaching their ideal. The Collings finally produced a bull named Favorite that met their requirements and immediately applied Bakewell's mating system of inbreeding. They mated Favorite to his daughters, granddaughters, and occasionally to his female descendants in the fourth and fifth generations. Comet, an illustrious sire and the first Shorthorn to sell for five thousand dollars, came from the mating of Favorite and Phoenix, a heifer that had been produced from the union of Favorite with his own dam.[3] To the unfamiliar, this inbreeding may seem shocking because it increases the expression of recessive characteristics. Remember, however, that inbreeding increases the homozygosity of desirable genes as well as undesirable ones. If the breeder can withstand the losses associated with rapid inbreeding, he ends up with the desirable traits he selected fixed in his herd. These desirable characteristics will be passed on to the offspring of the herd because the population is homozygous for them. The Tees Valley cattle were a variable population that represented a portion of the greater population that made up the species. When the Colling brothers finished their inbreeding program, they had further circumscribed the genetic variability to form the basis for the breed called Shorthorns.[4]

The bloodlines Favorite imparted to the Colling Shorthorns showed an improvement in fleshing ability and a refinement in quality that soon drew the attention of contemporary breeders. In a shrewd piece of advertising, the Colling brothers fitted (i.e., washed, brushed, and curled the coat and polished the horns and hoofs) the celebrated "White Heifer That Traveled" and the thirty-four-hundred-pound Durham ox and showed them throughout the United Kingdom, focusing the attention of the British public on the new breed.

During the 1830s the center of Shorthorn development shifted from England to Scotland, where Amos Cruickshank, of Sitlyton, Aberdeenshire, developed his Champion of England bloodlines. The Scottish lines of Shorthorns were a vigorous, early-maturing, and well-fleshed breed that immediately became popular.[5]

The first Shorthorns to reach America were English or Colling-type Shorthorns brought to Virginia in 1783 by the firm of Gough and Miller. By 1800, cattle bred from these imports had penetrated to Ohio and Kentucky. After 1880 the Scottish Shorthorns rapidly increased in popularity in America. The Scottish type fit the need of the times for a vigorous, early-maturing animal that did not require a huge quantity of feed or a long feeding period to be finished for slaughter.[6]

Showrings at the major livestock exhibitions became the battle-grounds for the purebred cattle industry in America. The overweight, fitted, and pampered animals of the livestock shows that became the standards of the industry probably would not have survived if turned loose on the western range. The breeders were not interested in selling these show animals, though, except for an occasional herd bull sold to another breeder. The glitter of the showring and the plump, fitted animals were illusions created for the western rancher who journeyed east with cash in hand. The breeders had bulls to sell, and at premium prices, but not these bulls.

Improvement of the Spanish-type Longhorn began in Nevada in the 1870s. In 1874 Senator Gabriel Cohn and John M. Dorsey of Lamoille in Elko County purchased from Saxes of Kentucky recently imported Shorthorn bulls with a value of five thousand to twenty thousand dollars.[7] Their offspring were worth seven dollars to ten dollars per head more than straight Longhorn steers. Equally important, the heifers that resulted from mating the Shorthorn bulls and Longhorn cows proved to be excellent brood cows. The hybrid heifers were fertile, vigorous, good milkers, and produced large, vigorous calves. In sum, they expressed all the qualities associated with hybrid vigor.

There is much confusion in nineteenth-century accounts over the names Durham and Shorthorn. The origin of the name Durham has been

linked with the county of that name in the Tees Valley. American import-
ers of Shorthorns apparently attached considerable prestige to this geographical location.[8] Many people considered the all-red Durhams a distinct breed. In fact, they were Shorthorns; the breed included red, white,
and roan animals. James MacDonald was surprised to find that Americans preferred all-red Shorthorns in the 1870s and were paying a premium price for them, and he was eager to return to Scotland to export red
animals to America to take advantage of the situation.[9]

Shorthorns were not the only English breed imported to upgrade the
Longhorns. Galloways were running on the North Fork of the Humboldt.
The English-financed Nevada Land and Cattle Company imported
polled-back Angus bulls from Aberdeen, Scotland. For some reason this
affronted the Spanish cowboys on the ranch, who promptly roped and
castrated the expensive bulls.[10]

Although many other minor breeds were represented, Shorthorns
were the most numerous English breed on the nineteenth-century western range. At the turn of the century a second major breed suddenly became popular and soared in numbers to rank second behind Shorthorns.
This breed, the Hereford, had a predominantly red body with white
markings and a trademark white face. Herefords were destined to become extremely popular on the sagebrush/grasslands range. Although
they were among the first breeds introduced, Herefords were little
known in the United States until the opening of the range country. After
about 1880, Herefords grew rapidly in popularity until they were almost
supreme on the range.[11]

The earliest record of Hereford cattle is from the Wye River valley in
the country of Herefordshire, an area known for centuries for its good
grass and excellent beef. There is considerable evidence that the Hereford breed antedates the systematic selection and breeding of other pure
breeds of cattle in Great Britain. Benjamin Tompkins and William
Galliers Sr. were breeding cattle of the Hereford type before Robert
Bakewell began his historic improvement of the ancient English Long-

horns. Tompkins and others apparently possessed herds of a well-established type of Herefordshire cattle in the late eighteenth century, when Charles and Robert Colling were barely laying the foundation for the modern Shorthorn.[12]

The origin of the white face of the Hereford is shrouded in mystery. Some traditions credit the marking to Dutch cattle; others attribute it to cattle brought from Yorkshire. The white face was a fixed characteristic in many herds as early as 1788, although it did not become the universal standard for the breed until many years later. Pedigree registration for Herefords was not on a firm basis until 1878, when the English Herd Book Society was founded. In the years preceding this move, there had been much difference of opinion as to color of the breed, and the first two volumes of the herd book recorded mottle-faces, white faces, and grays.[13]

The first documented importation of Hereford cattle to America was in 1817 by Henry Clay of Kentucky. In 1840 the first aggressive efforts to establish the breed in America were undertaken by William H. Sotham, who in a partnership with Erastus Corning of New York imported twenty-two head. The Hereford became a significant breed in the United States in the 1870s when T. L. Miller of Beecher, Illinois, and Thomas Clark of Ohio began importing them. The Miller and Clark herds attracted more attention than any of their predecessors, and a number of breeders were drawn to the breed shortly before 1880. Among the herds founded at this time were those of C. M. Culbertson of Illinois, Fowler and Van Natta and Earl and Stuart of Indiana, and Gudgell and Simpson of Missouri.

Of the American foundation herds that influenced the development of the breed, none was more important than that of C. J. Gudgell and T. A. "Governor" Simpson of Pleasant Hill, Missouri. Gudgell, a banker, and Simpson, a farmer and horse dealer, formed a partnership to raise Shorthorns. In 1876 Gudgell and Simpson visited the Centennial Exposition in Philadelphia and saw Herefords for the first time. They liked what they saw, except that Herefords were generally too light in the hindquarters. Later, they saw a bull calf imported from England by C. M.

Culbertson of Newman, Illinois. This bull, Anxiety 2238, had the hind-quarters that Gudgell and Simpson believed the breed required. Unfortunately, Anxiety 2238 died before he had a chance to influence the developing American strain of Herefords. Still searching for a foundation sire, Simpson went to England to find a relative of Anxiety 2238. As Simpson prepared to leave America, his partner gave him some parting advice: "If you find a bull with an end on him, bring him with you." The bull Simpson returned with, Anxiety 4th, was thus known as "the bull with an end."[14] Anxiety 4th became a great herd bull and helped establish Gudgell and Simpson's Hereford herd as one of the best. Anxiety 4th died in 1890, but his offspring contributed to Hereford herds throughout the Midwest and in increasing numbers on the western range.

Not only were Hereford bulls popular on western rangelands, whole herds of purebred Herefords soon became established on the range as well. George T. Morgan, an Englishman, came to Wyoming in 1876 with the idea of introducing breeding stock and returned in 1878 with a shipment of Hereford bulls consigned to A. H. Swan. These bulls, reportedly the first in Wyoming, cost more than ten thousand dollars to import. With Morgan as his manager, Swan established the Wyoming Hereford Ranch on forty thousand fenced acres. Through the mid-1880s the ranch imported more than five hundred Herefords from England.[15]

Joseph Scott of the 71 Ranch at Halleck is usually credited as the first to import Herefords to Nevada. He imported twenty heifers and four bulls in the late 1870s. Scott, born in Ireland of Scottish parents, developed ranches in Montana, eastern Washington, and Wyoming before coming to Nevada. He came to prominence in the purebred Hereford business in the firm of Scott and Hank of Mandel, Wyoming, which bought foundation stock from C. M. Culbertson and also imported stock from England.[16] In 1880 Scott and Hank were taxed on the Elko County tax rolls for fourteen head of purebred Herefords valued at one hundred dollars each. By 1886 they had sufficient purebred stock to sell a yearling to Russell and Bradley for four hundred dollars and a cow and calf to Williams and Hunter for one thousand dollars.[17] Another early Nevada leader in importing high-quality bulls was Abner Cleveland, cousin of the U.S. president.[18]

The purchase of quality bulls to improve their herds involved a considerable capital investment for ranchers. Prominent rancher and later governor Jewett Adams used a unique method of raising capital to purchase a carload of whiteface bulls. He won thirty thousand dollars in a high-stakes poker game and promptly invested it in Herefords.[19]

In 1885 the Elko newspaper announced that John Sparks of the giant ranching firm of Sparks-Tinnin was leaving Elko County, where he had resided since 1881 on the remote H-D Ranch, for the western Nevada city of Reno. Although he credited the hot springs on the ranch with restoring his wife's health, he had decided to move on. The reason he gave for the move was to improve educational opportunities for his children.[20] Sparks continued to operate the extensive ranching properties in northeastern Nevada and south-central Idaho after he moved to Reno. He also maintained his residence in Georgetown, Texas.

In 1887 Sparks made another significant change. He purchased Anderson's Station, 1,640 acres south of Reno on the former site of the Junction House, a way station that lay in the center of the major immigrant trails passing through the Truckee Meadows area.[21] Sparks, who always had a flair for the colorful, named his new property Alamo and constructed a striking plantation-style residence. The Alamo Stock Farm was located in the geothermal belt north of Steamboat Hot Springs, and two artesian wells were located on the property. One well tapped hot water 240 feet below the surface, and the second penetrated to a depth of 560 feet. The hot water was funneled into a large swimming pool.[22] Perhaps Mrs. Sparks did not miss the hot springs of the H-D Ranch.

Later Sparks purchased the four-hundred-acre Mayberry Ranch on the Truckee River west of Reno, increasing the irrigated pasture and hay land under his control in the Truckee Meadows area. The Virginia City and Truckee (V & T) Railroad tracks bisected the Alamo Stock Farm and for Sparks's convenience provided a stopping point at his headquarters. Sparks set about making his Truckee Meadows property into a showplace. He imported American bison, elk, Persian sheep, and various types of deer. Passengers on the V & T were treated to a parklike atmosphere as they rode through the Alamo property.

The most important animals John Sparks purchased to stock his Truckee Meadows ranches were purebred Hereford breeding stock. After the turn of the century, Sparks told an interviewer that he started in the registered Hereford business in 1875. He probably meant he started experimenting with Hereford bulls for crossbreeding Longhorn cows at that time; apparently, his first attempts to raise his own purebred animals did not occur until after he developed the Alamo ranch.[23]

In the late nineteenth century the *Pacific Rural Press* included a gossipy information column for livestock buyers. The January 5, 1889, column noted that John Sparks of Reno, Nevada, had five hundred steers on feed that winter. The same issue indicated that John Slaver, a Truckee Meadows livestock feeder, had shipped five railcars of fat steers to Grayson, Owen and Company in Oakland, California. The steers had been purchased from the ranches of Sparks-Tinnin and included several thoroughbreds as well as half-breed and three-quarters-breed Hereford steers.[24]

There are two methods for developing a successful herd of purebred animals. The prospective breeder can either purchase a fair number of females of the general type he desires and breed them to a quality foundation sire, select the offspring based on the standards of quality he wants to establish, and embark on a controlled inbreeding program to fix these qualities in his herd; or he can go to showring sales and spend large amounts of money for flashy, established-name animals and hope for the best when they are mated. Essentially, the second method involves purchasing from outbreeding populations in contrast to inbred populations. With his ready cash and flair for publicity, Sparks chose the second method. Most of his noted national and international purchases were made in the 1890s after the range operations were reorganized as Sparks-Harrell.

In 1894 Sparks purchased Hereford breeding stock from C. H. Elmendorf of Kearney, Nebraska. This herd made impressive winnings on the stock-show circuit using animals that were the progeny of the foundation bulls, Autocrat and Earl of Shadeland 3oth. Before that, Sparks had visited Missouri and purchased animals from Fielding Smith and even from the foundation herd of Gudgell and Simpson.[25]

Despite the fact that the livestock industry on the sagebrush/grass-lands was still recovering from the disaster of 1889–90, and even though the country as a whole and Nevada in particular were in a deep economic depression, John Sparks had money to spend in the early 1890s. He purchased almost the entire first-calf heifer crop of the Wyoming Hereford Ranch in 1895 from his old associate of the beef-bonanza days, Alex Swan.[26] These heifers were bred to Luminary 81654, Beau Brilliant 86753, and Patrolman 91594, all quality sires.

In one of their last sales, Gudgell and Simpson and another outstanding Missouri Hereford breeder, James Funkhouser, sold ninety-seven animals at auction for a record average price of $278 per head. John Sparks of Reno, Nevada, was the liberal buyer.[27] Missouri auctioneers must have rubbed their hands with glee when they had John Sparks at the sales ring. Tall, handsome, dressed expensively in western clothes with a gun under his coat, Sparks himself was an attraction. The national livestock journals, *Breeders Gazette* and *Weekly Drover's Journal,* had both carried stories calling Sparks "the largest rancher in the West." When the auctioneer turned to Sparks for one more raise on a high-priced bull, the crowd held its breath. How could he refuse, even if he was raising his own bid?

During the late 1890s John Sparks ran at least one advertisement in each issue of the *Reno Evening Gazette* offering his registered Hereford bulls for sale. Not surprisingly, he enjoyed a very favorable press. In 1898 the editor urged local Nevada ranchers to buy the surplus Hereford bulls that the Alamo Stock Farm had not been able to sell.[28] Sparks was an innovator in the marketing of registered Herefords. In 1896, for example, he employed the most popular photographer in Reno, E. P. Butler, to take photographs of his prize Herefords. Photographs were not reproduced in newspapers at that time, but the story of a photographic session for bulls was wildly reprinted in Nevada papers.[29]

Sparks loved to take visitors on tours of the Alamo Stock Farm to show off his collection of exotic animals and, more important, his registered Herefords. The interior of the Sparks mansion was furnished with many curios of his life on the frontier which he also enjoyed showing to visitors. He once paid one hundred dollars for a set of deer antlers locked together in fatal combat. A servant was assigned the task of polishing the antlers. The overzealous servant returned to report to Mr. Sparks that he had had a terrible time, but he finally succeeded in separating the antlers.[30] After entertaining his guests with curios and stories, Sparks refreshed them with glasses of his special applejack. The editor of the *Reno Evening Gazette* claimed that one drink of Sparks's applejack would make hair grow on a bald head within twenty-four hours.[31] By 1897 the same paper had reduced the time necessary for hair to sprout to fifteen minutes.[32]

John Sparks was very good for the developing Hereford industry. He was no dull midwestern farmer who raised fancy English cattle in a barn; he was the king of the western range, the biggest of the big. A story soon began to circulate in the livestock industry that Sparks started breeding Herefords because he discovered that they had survived the winter of 1889–90 much better than Longhorns. Reinhold Sadler, one of the cattleman governors of Nevada, even included this story in a speech to his constituents.[33] The governor reported that Sparks was running a herd of sixty-five thousand in 1889 and lost thirty-five thousand. Of the survivors, 90 percent were Herefords. This kind of publicity was pure gold for Hereford breeders.

A compilation of historic facts about Herefords in America quotes

Sparks as saying in regard to the winter of 1889–90, "I lost about 30,000 cattle or about 65 percent of my entire herd. The Herefords at the beginning constituted about 40 percent of the whole herd. I found that of the number surviving the second winter, at least 90 percent were Herefords, showing conclusively their superior constitution."[34] This statement must have resulted in hundreds of thousands of dollars in bull sales on the western range. Possibly the most accurate direct quote from Sparks on the losses of 1889–90 was published by *Harper's Weekly*: "We lost that winter, which was a severe one, 35,000 head of cattle and when we rounded up our cattle the following spring, 90 percent of those we found had white faces characteristic of Herefords."[35]

The numbers may be questioned, but the vital point of this quote is the mention of the *white faces* characteristic of Herefords. The animals that survived were probably hybrid offspring of Hereford bulls and Longhorn cows. Their survival was a product of hybrid vigor. The Hereford bulls contributed to this vigor, of course, but so did the genetically diverse Longhorn cows.

The Sparks mansion contained two cases of trophies and medals won by his Hereford show herd. Sparks fitted a show herd that toured state fairs and shows all over the far West. Articles about Sparks often mention Earl of Shadeland 30th, the Alamo herd's top bull, as the champion of the Columbia Exposition of 1893.[36] The bull stood third in the sweepstakes class for bulls of any age. In fact, at the time of the exposition, Earl of Shadeland belonged to C. H. Elmendorf. Sparks bought the bull later and gradually started referring to him as "his" champion of the Columbia Exposition. Regardless of who owned him, Earl of Shadeland 30th was a tremendous bull who for several seasons was considered unbeatable on the midwestern show circuit.[37]

John Sparks received a lot of publicity from Earl of Shadeland 30th, but it was purchased publicity, and not a product of his own breeding program. Apparently, the best finish in national competition for breeding stock obtained by progeny of his own program was a fourth-place finish by a junior yearling bull in the St. Louis World's Fair of 1904.[38]

During the late nineteenth century there was considerable prestige attached to importing Herefords directly from Herefordshire, despite the growth and progress made by American breeders. From 1880 to 1900 some five thousand head were exported to the United States. This must have left English pastures considerably depleted of stock while enriching English Hereford breeders. Sparks imported eight Herefords, two bulls, and six heifers during this period. One of the bulls was purchased from the Monkton herd of James Smith at Pembridge, Hereford, one of the first established herds of Herefords in England. The second bull was purchased from the Courthouse herd of John Price, also at Pembridge, Hereford. This herd was founded in 1862 and had a brilliant show record in England. The heifers were purchased from A. P. Turner's Leen herd and R. Green's Whittern herd as well as the herds of Smith and Price.[39] These eight animals were sufficient to establish Sparks's image as an importer. One exuberant supporter said that Sparks's herd "was capable of competing with the royal herd of Queen Victoria."[40]

Sparks was quoted after the turn of the century as saying that he began importing Herefords in 1893.[41] Apparently, however, his first venture in English cattle occurred in 1897 when he visited Kansas City to see a shipment of Herefords imported from England by Kirkland Armour. He hired George Morgan, a native of Herefordshire who had helped deliver the Herefords to Armour, to manage his Alamo herd. In 1900 Sparks sent Morgan back to Herefordshire with a bank draft for ten thousand dollars to purchase a quality herd bull.

When President Theodore Roosevelt visited the Alamo Stock Farm in 1904, Sparks told him, "Now that you've met my prize bulls, Mr. President, meet my John Bull," and introduced George Morgan. "If you want to ask any questions about the bulls," he continued, "ask George."[42] This may well have been a tacit admission that review of long pedigrees and reciting English bloodlines was not among the many talents of this cowboy capitalist from Texas. Sparks may have lacked technical details in the purebred livestock business, but it did not keep him from being active in the politics of the business. He was an active supporter and early president of the American Hereford Association.[43]

John Sparks reached out for technical expertise by hiring William Stevenson, whose father had been the manager for Gudgell and Simpson. The Stevensons were Scottish cattlemen of the old school. Stevenson became the manager of the Mayberry Ranch in Sparks's Truckee Meadows holdings.[44] In 1900 John Sparks capped his national and international purchases of Herefords by paying the unheard-of price of ten thousand dollars for Dale 66481, one of the top sires in the national show circuit.[45]

Always quick to grasp the potential publicity value of his actions, John Sparks protested the values the Washoe County tax assessor placed on his purebred Herefords in 1901. The assessor valued his bulls at seventy dollars and his cows at fifty. Sparks raised the assessed value to five hundred dollars for the bulls and one hundred dollars for the cows. Sparks appeared before the Board of Equalization and refused to pay until the valuations were raised.[46] He succeeded in getting the assessed value of his bulls, which were his marketable product, increased to more than seven times the original amount, while the cows, which made up the bulk of his herd on a continuing basis, only doubled in assessed value.

In 1889 Sparks became involved in a national scheme to promote Herefords. Kirkland Armour donated a Hereford heifer to the promoters of the American Royal Livestock show in Kansas City to be raffled off to raise funds for a building program. This was not just any heifer, but Armour Rose 75084, a perfect yearling. She was a product of inbred Anxiety blood, the get of Beau Brummel Jr., bred by Gudgell and Simpson. Thousands of raffle tickets were sold. A lady held the lucky ticket. Armour bought the heifer back for one thousand dollars and redonated the animal. The heifer was bought and redonated several times until finally John Sparks and the manager of Marshal Fields's Hereford herd at Madison, Nebraska, got into a spirited bidding war. When the auctioneer's gavel fell for the last time, Sparks had purchased Armour Rose for the then-record price for a Hereford female of twenty-five hundred dollars. Armour Rose's dam was also related to Beau, and the inbreeding was too close; she was barren and never produced a calf.[47] Although this may appear to be a net loss for Sparks, the national publicity he received added to both the number of bulls he sold and the price he received for them.

Sparks proved he could still recognize quality when he sent a steer named Alamo to the Chicago International. The steer was named grand champion at this show, an award that previously had been the exclusive province of Corn Belt stock feeders. The Armour Packing Company killed the steer, which dressed an unbelievable 70.1 percent. Sparks knew when to keep a good thing going. He had Alamo's hide tanned and presented his mounted head to Thomas Marlow, president of the First National Bank of Helena, Montana.[48]

The fortunes of the Alamo Stock Farm were on an upswing in 1899 when James Wilson, secretary of agriculture for President McKinley, visited Sparks's operation.[49] Sparks sent his show string of Herefords to the Kansas City Royal, which hosted the greatest show of Herefords to date. More than five hundred highly fitted animals were exhibited, and three hundred head were sold at auction at an average price of $317.[50] Sparks returned to Reno to announce that his stock were the highest-selling animals at the show.[51] By the turn of the century America had become an exporter of purebred livestock. Sparks became the first Hereford breeder to ship purebred animals to Honolulu, Hawaii.[52]

Sparks's purebred Hereford operation influenced livestock production on the sagebrush/grasslands in three major ways. First, by supplying quality bulls to ranchers and assembling a critical mass of quality purebred Herefords, Sparks provided a foundation for Hereford breeders that had an influence long after the Alamo Stock Farm was dispersed. Second, Sparks's national reputation enhanced the image of Intermountain agriculture. The third point is obvious from the previously mentioned visits of the secretary of agriculture and President Roosevelt to the Alamo Stock Farm. Only half a century before, the Intermountain area had been known as the dreaded desert blocking the way to California and Oregon. Its image had changed enormously.

The Alamo herd provided foundation stock for the Whitaker herd of Galt, California; the Jack herd of Saunas, California; and Joseph Marsden of Lovelock, Nevada.[53] Many of the top bulls from the Alamo Stock Farm went to Cazer and Son of Wells, Nevada. Prospective Hereford breeders in what had once been the most remote backwaters of the United States suddenly had access to bloodlines of the most valuable Herefords in the world.

During the 1890s, environmental constraints limited the ability of ranchers to meet U.S. consumers' demand for smaller cuts of tender beef. Midwestern feeders were demanding a feeder animal from the western range that matured early and was an economical converter of feed to meat. The term "beefs," meaning huge steers that matured at five years old, disappeared from the livestock industry. Ranchers in the sage-brush/grasslands had few options available to them. Depleted sagebrush ranges cannot produce sufficient forage to fatten two-year-old steers. Only wetlands and meadows produced forage of quality and quantity to fatten beef, and meadows were in extremely short supply in the Intermountain area. Even if ranchers had the physical environment, irrigated land, and natural meadows on which to finish beef, they still needed to produce an earlier-maturing animal through breeding.

Just as the "White Heifer That Traveled" helped establish Shorthorns as a breed in eighteenth-century England, livestock shows were the means by which new concepts and trends of breed quality were imparted to nineteenth-century American ranchers. Early maturity became the keynote of the stock shows in both breeding and fat classes. In the late nineteenth century, two calves won the grand championship at the Chicago International over fat steers of all ages, emphasizing the tendency toward smaller and younger beef and eliminating age as a barrier to the fitness of a carcass for slaughter.[54]

The thrifty English breeds of cattle provided the genetic material to change the characteristics of the Longhorns. Coupled with the development of refrigeration, the new type of cattle turned twentieth-century Americans into confirmed beef eaters, and a thick, juicy steak into the symbol of quality. An analysis of the process by which this occurred reveals a bizarre sequence of events. How could animals developed under the humid environmental conditions of England, where pastures consisted of a hedge-rimmed acre and animals were kept in barns during winter, become adapted to the arid, fenceless sagebrush/grasslands?

Even after the English breeds were introduced to America, the standards of quality were based on performance in showrings rather than performance on the range. Much of the success of the English breeds was due to the linebreeding system. Breeds originated because herdsmen followed the example of Robert Bakewell and fixed desirable characteristics through inbreeding. Within established breeds, certain breeders continued linebreeding to fix certain bloodlines within the breed. The Anxiety Herefords are a prime example of this technique. Breeders persisted with linebreeding when the advantages were seen in the progeny. The payoff for this continued inbreeding occurred when the English breeds were crossed with the Longhorn cows. The resulting offspring expressed heterosis. This hybrid vigor meant that the animals could survive winters like that of 1889–90, be more fertile, and mature earlier. Without the Longhorns, the system would never have worked.

One outgrowth of the introduction of English breeds was the equating of quality with uniformity. Obviously, the product was of greater value to feeders if the steers produced from the sagebrush ranges were of uniform size and conformation. But this uniformity was extended to color, horn shape, and other characteristics that contributed little or nothing to carcass quality or feed-conversion efficiency.

Unfortunately, the hybrid vigor declined with each generation that improved bulls were used. The first time Herefords were crossed with Longhorns, the response seen in the offspring was tremendous. When the first cross of offspring was bred back to Hereford bulls, the expression of hybrid vigor exhibited by their offspring was proportionately less with each generation. The range herds gradually lost the appearance of Longhorns and assumed the coloring and appearance of the English breeds. All these interacting factors would have a tremendous influence on twentieth-century ranching in the sagebrush/grasslands.

Ranch technology changed enormously between 1860 and 1900. At the beginning of that period Longhorns in Texas were free-roaming, free-breeding populations with virtually no interference from man. By the turn of the century, ranchers were investing scarce capital in English bulls and worrying about linebreeding and bloodlines while trying to produce baby beef. At the end of the Civil War, John Sparks was a cowboy driving wild Longhorns up the virgin grasslands of the Great Plains. Ranchers had no capital improvements, and the cowboy's tools were restricted to horse, saddle, and rope. By the end of the century, John Sparks was hiring imported English herdsmen to give purebred Herefords a bath and curl their hair before parading them in a showring!

CHAPTER 12

Horses, Tame and Wild in the Sagebrush

The horse was a necessary part of livestock management on the range. Without the horse, the culture of cattle on the sagebrush/grasslands would have been impossible. Of all the domestic livestock introduced to the sagebrush/grasslands, only horses and burros became successful feral animals, although one could argue that camels were never given a fair chance.[1]

Horses evolved in North America, but all the native horses were long extinct by the time domestic horses were introduced into the Western Hemisphere. The North American evolution of the horse is interesting for several reasons. Although the sagebrush/grasslands and other vegetation formations of the semiarid and desert West are of comparatively recent origin, at least portions of the vegetation evolved from formations that existed and evolved under grazing by the ancestors of modern horses and other large herbivores.[2] Some aspects of the modern sagebrush species—such as the essential oil content of the herbage that inhibits browsing—may have been of adaptive value and become fixed in the gene pool of the ancestor population when the plains of the West were browsed by herds of native herbivores.[3] A second point of interest concerning the evolution of horses in North America is the disappearance of most of our native large herbivores in sudden mass extinctions at the close of the Pleistocene.[4] This sudden extinction is especially puzzling in the case of horses because of the obvious success of feral horses in western North America.

The true wild horse, the remote ancestor of the domesticated and feral horse of our era, evolved on the American continent at a time when semitropical woodlands and lush vegetation provided a hospitable environment. The first horse was about the size of a fox and had four toes on the forefoot and three on the back. Thousands of crushed skulls and skeleton fragments of this tiny animal, called *Eohippus*, have been found in rock formations of the West.[5] About fifty million years ago *Eohippus* grazed over the area that is now Wyoming, New Mexico, and Utah. Many species of *Eohippus* developed during the millions of years of the Eocene epoch. Some died out; others developed modifications over the years that made them more successful. By the Oligocene epoch, some species had lost one toe and begun to depend on the center digit more than the outer two; the facial portion of the skull had become slightly longer than the cranium. The brain had become more complex, which probably made them much more cunning animals. These creatures, known as *Mesohippus*, were six to eight inches taller than *Eohippus*.[6]

During the Miocene epoch, some species developed from *Mesohippus* that were better adapted to open grazing rather than to the forest environment. The coarse plains grasses with their high silicon content were hard on teeth, and during the millions of years of the Miocene epoch the more successful grazers developed teeth with long crowns and enamel ridges with cement in between, much like the horse teeth of today. The skull length increased to facilitate grazing, and the dependency on the middle toe increased as well. Ultimately, the most successful equine species of this epoch turned out to be *Meryohippus*.

By the middle of the Pliocene epoch, some one-toed species had evolved. The protohorses increased in size throughout the millions of years of this epoch, so that finally *Pliohippus* resembled some of our zebras of today.

Equus was the typical equine of the Pleistocene epoch. No one knows how many species of this genus evolved and died out in the millennia during and after the Pleistocene. Modern members of the family Equidae are divided into nine species. There are two species of horses: *Equus caballus* and *E. przewalskii.* There are probably four species of asses and three definite species of zebras.[7]

During the Pleistocene, when much of the oceans' water was tied up in glacial ice, a land bridge existed between North America and Asia across the Bering Strait. The horses that had evolved in North America spread north and westward across this bridge and populated Asia, Europe, and Africa. If not for this movement we might now be in a horseless age, because all of the native horses became extinct in North America after humans crossed the same land bridge from Asia to North America.

The modern horse probably evolved in Central Asia. There is much evidence that modern horses of the species *E. caballus* developed along two lines often referred to as heavy horses and light horses, recognizable by their size and proportions. The light horse had a short face, a narrow snout, and slender legs and body. The heavy horse had a long, narrow skull with a prominent face and massive bones in the body and legs. There are differences in enamel patterns of the teeth of the two types as well.

Columbus deserves the credit for introducing the modern horse to the Western Hemisphere. On his first trip he was instructed to take 6 mares, 4 jackasses, 2 she-asses, 4 bull calves, 2 heifers, 100 sheep and goats, 80 boars, and 20 sows.[8] Subsequent voyages greatly increased the horse population of the New World. So many horses were taken from Spain, in fact, that in 1505 the king of Spain forbade further exports to allow the domestic horse herds to increase.

The horses imported into North America are thought to have been

Andalusians bred on the plains of Córdoba. The Andalusian is an old Spanish breed that dates to the Moorish invasions in the eighth century A.D., when Barbs and Arabs brought by the Moors were crossed with the local Spanish ponies and later with larger horses. Certainly the Moors were a prominent influence in the breeding of Spanish horses. The Andalusian has been important in the development of many other breeds, partly because of the influence Spain has had on other cultures. Andalusians played a role in establishing the Criollo and Campolino breeds of South America as well as the mustang in North America.[9]

The northward movement of horses in North America seems to have been along two lines, one paralleling the eastern slope of the Rockies, the other west of the Continental Divide.[10] The Snake or Shoshone Indians apparently were the brokers through whom the tribes of the Northwest gained horses. The Shoshone probably had horses by 1700.[11]

There were thousands of wild horses in California during the Mission period. They were so numerous, in fact, that during periods of drought they were killed to provide more forage for cattle.[12] Quite possibly the first reported sighting of wild horses in Nevada was made by John Bidwell, who saw a solitary horse near the sink of the Humboldt in 1841.[13] Dan DeQuille and his party saw seven wild horses in the Stillwater Range while prospecting in the summer of 1861.[14]

The Indians native to the central Great Basin were not, at least initially, great horsemen. By most accounts, horses were relatively scarce in most of northern Nevada. The economy of the endemic Indians was based on small family units of seed gatherers and rodent hunters. Trade was limited, and the general economy was in such a state that warfare was limited as well. No one had anything worth fighting for, and mere survival was of paramount importance. Under such conditions, the occasional horses seen in the area were viewed as valuable food rather than transport and were quickly dispatched by rabbit hunters lucky enough to capture them.[15]

The early European explorers in central Nevada documented the Indians' lack of familiarity with horses. For example, Captain J. H. Simpson, who explored central Nevada in 1859, reported the attempt of a Paiute to mount a mule. "In mounting his mule, he invariably would protrude his legs through and between his arms while resting his hands on the saddle, and, in one instance, in his attempt to mount this way, awkwardly tumbled off the otherside."[16] It did not take long, however, for the Great Basin Indians to become horsemen avid for more animals. The history of the California Trail along the Humboldt and the Oregon Trail across the Snake River Plains is profuse with cases of Indians stealing or trying to steal horses.[17]

Horse ranching quickly became very profitable in the sagebrush/grasslands. I. V. Butlon started ranching in Humboldt County, Nevada, in the early 1870s. His Double Square Ranch specialized in horses. He sold remounts to the U.S. Army and police horses to the city of San Francisco.[18] Frank Fernald specialized in raising draft animals for freight lines, which demanded a constant supply of workhorses. He sold heavy draft horses for $150–175. Thousands of workhorses were raised along the Humboldt River. One ranch had more than a thousand head.[19] Southeastern Oregon and northern Nevada were well known by 1880 for producing fine cow horses. Untrained horses sold for $60 in 1880; trained horses often sold for $100.[20] The McIntosh brothers of Carlin, Elko County, Nevada, specialized in raising trotting horses, purchasing the stallion Whippleton Jr. as their founder.[21]

Draft horses were an important part of late-nineteenth-century agriculture and the general economy. The Clydesdales that appear on television commercials give some idea of the grandeur of a fine team of draft horses prancing down the street with haines bells ringing. Only those who have seen them in person, however, have a true appreciation for the breed. Draft horses are big. Light draft horses weigh 1,200–1,400 pounds; medium draft horses, 1,400–1,600 pounds; and heavy horses,

1,600–2,000 pounds. The major breeds include Percherons or Normans from France, and Shires, Clydesdales, and Belgians. E. C. Hardy of the Oasis Ranch near Toano, Nevada, imported quality Norman Clydesdale stallions and mares from the East and became a prominent supplier of draft horses in the central Great Basin.[22]

We have already alluded to the tremendous number of horses required for hay making in the sagebrush/grasslands. In the fall, ranchers who were good managers would do their necessary tillage work with the horses that had been hardened into good shape by the haying work. It was more efficient than waiting until spring when the horses were out of shape.[23]

When haying and tillage operations were finished, horses were turned loose on the range to rustle for themselves. In midwinter the horses might be fed supplemental straw, but generally they were dependent on native grass until the next hay season. The winter survival abilities of horses must be one of the keys to their success as feral animals. Horses climbed up to the high ridges where the snow blows free and worked the snow-free south slopes. They trampled the bunchgrass clumps on the steep slopes when the soil was wet and ate the perennial grasses in the early spring when the grasses first started to green up. This type of horse operation was hard on the range, and it was hard on the horses, too. Many ranchers treated horses as an expendable resource cheaper to renew than to give unnecessary care.

Almost every ranch had bands of free-roaming horses running on its range. Each band consisted of a dominant stallion, mares, and tag-along geldings. The horses used on the ranch passed in and out of these bands during the off-season. Ranchers often deliberately turned loose quality stallions to breed up these bands after shooting or trapping the free-roaming stallions.

What were these free-roaming bands of horses like? Frank Dobie attributes some of the most realistic descriptions of mustang bands and mustang trapping to the writings of Rufus Steele, and indeed, Steele's words evoke a clear picture of the breed.[24] "You may be riding along carefully among towering mountains," Steele wrote,

> when, quite suddenly, you come upon a herd of wild horses feeding or standing half asleep in the shade of rocks or stunted trees. One of the herd sees, hears, or smells you, and instantly all are alert. If you rein in your horse and remain motionless, the wild stallion will advance toward you with extreme caution. At last he halts, throws up his head, emits a mighty snort, and instantly he is away at full speed with his herd at his heels. Down the mountainside they go with never a trail to follow. They leap, scramble, tumble, and crash through old dead timber, and when they strike a bit of good running ground, their hoofbeats come back to you like the roll of a drum. If they are pursued, the thick-necked, thin-legged, many-scarred stallion continues to lead. If no pursuer appears, the stallion drops to the rear, to be on the alert against surprise, and his place in the lead is taken by a crafty old mare. During long runs, I have witnessed this change in leadership many times.[25]

Capturing wild horses was no easy matter. Even a cowboy who had worked around mustangs for years, was an expert and fearless rider and a sure roper, knew the range perfectly, had toiled unremittingly in his preparations, and was assisted by men of equal experience might be outwitted by a wily stallion or a sagacious old mare. One method of capturing wild horses was to run herds with a relay of riders. When the country was sufficiently open and level, five or six experienced men, well mounted and properly stationed, could sometimes keep a herd of horses running in great circles. By relieving each other at regular intervals, they could, in time, wear out the wild horses and corral those that did not give out during the run.

The distances these horses could run when thus pursued by relays of riders are almost beyond belief. Bands that had been run for twenty miles could turn on a spurt of speed and outrun fresh horses. Bands that had been chased a few times discovered that the pursuers were not after individuals and learned to drop away from the herd one at a time and escape. At length the pursuers would find themselves trailing only one or two horses and give up in disgust.[26]

A band of horses that started to run would race away for a short distance, then halt and face about at the crest of the first ridge, like a line of soldiers. If they saw the pursuer coming, they snorted, wheeled about, and started on the long, long race. It was up to the mustanger chasing them to keep crowding the horses and to keep them running in the correct direction. Often this resulted in head-to-head confrontations between the mustanger and the stallion leading the band. Mustanger Pete Barnum often rode neck-and-neck with game stallions, beating them across the face with his quirt until their faces were drenched with blood, only to have the stallions slacken their speed, dodge behind his horse, and continue on their contrary way.

Running wild horses was hard work for horses and mustangers alike. Mel Sharp, who grew up running mustangs in Nye County, Nevada, claimed that if he did not have his saddle blanket under his arm at the end of the run he was in trouble with his father for dogging the drive. When the mustanger started the run on a fresh horse, the saddle cinch would be tight. As the run continued, the horse would start to work out from under the saddle. There was no time to stop and tighten the cinch, so the rider was forced to hold the saddle blanket under one arm, keep the saddle on the horse with his knees, hold the reins in his teeth, and whip his tired horse with a quirt held in his free hand.[27]

As a boy, Tom Bunch was running horses in the Snake River breaks of Baker County, Oregon, and took a break to stop at an abandoned homestead to pick green apples. By tightening his belt and tucking in his blue denim shirt, he was able to pack half a bushel of hard green apples around his middle—just the thing to throw at horses! Unfortunately, in his haste to rejoin the run he did not see a squirrel hole, and his horse cartwheeled at a dead run.[28]

A less strenuous method of capturing mustangs was to drive a bunch of gentle horses onto the range. Bands of mustangs were then worked into the gentle, or *parada*, band. Theoretically, the gentle horses calmed the mustangs and the entire band could be driven to corrals.

Pete Barnum, whose mustanging activities were recorded by Rufus Steele, revolutionized the art of trapping wild horses by inventing the canvas corral. After noticing that wild horses seldom jump barriers that they cannot see through or over, he developed canvas corrals fifty feet in diameter and seven feet high. He used cloth bunting wings to direct the horses into the trap. The portable trap was light enough to be packed on mules and could be set up in two hours.[29]

In the more arid portion of the sagebrush/grasslands in southern Utah and Nevada, water hole traps were used to trap mustangs. Steve Pellegrini described mustangs coming to water in the mountains west of Walker Lake:

In the summer, the horses of the Wassuk Range typically return to their water hole once every other day. Although the favorite time is just after dark in the evening, they may also come in during pre-dawn. They water individually, probably due to the small spring size in the Wassuks; the oldest mare and her colt water first and the band stud last. It takes each horse about five minutes to drink, and if the stud tries to water before his turn, the oldest mare usually drives him back. In about an hour, each animal has time to water again and the band settles down to spend the night at the water hole. They roll, sleep, and do some feeding, but usually do not go more than one hundred yards from the spring. They also do much pawing at the water, a habit which keeps the mud from filling it in.[30]

Corrals were built around watering points with gates that could be easily shut from blinds. At first, the free-roaming horses refused to approach such traps, but the limited number of watering points soon forced them to come in. This process could be speeded up by closing access to all the neighboring springs.

Wild horses are extremely difficult to handle, and even with the best of care, some were bound to injure themselves when trapped. Pete Barnum roped the horses he wanted out of his canvas corrals, threw each down, and bound up one front leg firmly at the joint. When the horse was released, it would spring up on three legs and try to run. The horses were driven to the breaking corrals with a painful three-legged gait. Around the sagebrush/grasslands several variations of this disabling technique were used to handle rough horses. In a later period, the Utah Construction Company used a "W" made of cotton rope tied around and between the horse's front legs. A cruder method consisted of tying one of the horse's legs to its tail.[31] Some mustangers used the damper method of controlling the horses they caught, accomplished by cutting a small slit in the outer side of each nostril. A small buckskin strip threaded through these slits could be drawn up to partially to close off the air. This regulated the speed of the horse and kept him under control during the drive from the trap to the home ranch. The buckskin string became moist from the moisture of the horse's breath and after a few days stretched to allow the nostrils to expand. Some mustangers cut the slits for the buckskin strings very thin so the strings would tear out in a few days. Although this treatment sounds cruel, old-time mustangers defend it as less dangerous to the horse than tying up a leg. The horse with its nose sewn up could eat and drink and did not have to be roped again to remove a hobble or a tied-up leg.[32]

Pete Barnum employed Shoshone riders in his Nevada mustang operation. For breaking mustangs he used his canvas corral. An Indian roper caught the horse by its front feet and pulled it down. A second Indian jumped on the horse's head and held it down while a third roped the hind feet. Some bronc busters would jump on the mustang's head and firmly grasp one of its ears with their teeth. A snaffle bit was worked into the horse's mouth, and the horse was saddled while on the ground by working a rope to pull the cinch under the neck and in the back of the shoulder. A leather blindfold hung over its eyes calmed the horse while all this was taking place. Once the horse had been saddled, the bronc rider mounted the saddle and tried to catch the off stirrup as the horse rose. With raking spurs and a heavy quirt, the rider roughed up the horse while desperately trying to remain in the saddle. It might take two or three such rides before the horse could be saddled standing.

During the recurring periods of depressed livestock prices, out-of-work cowboys would run mustangs. Likewise, if a ranch hand had a misunderstanding with a foreman or ranch owner and went down the road, he always knew that he could go to running mustangs until another job turned up. Because of these factors, mustang populations tended to inversely mirror economic conditions. When times were hard, men preyed on the mustangs; when everyone was employed, the mustangs roamed freely. There were three notable exceptions to this general rule: (1) foreign wars, such as the Boer War late in the nineteenth century, created sudden demands for horses; (2) local mining strikes and rushes meant sudden predation on surrounding horse populations as the demand for transportation surged; and (3) local ranchers who wanted to raise quality horses kept continued pressure on free-roaming horse populations, especially stallions.

Mustanging approached being a disease with some men. More than one wife of a sagebrush cowboy had to put up with a husband who had periodic outbreaks of mustanging.[33] What was the attraction? It is difficult to pick a single factor that provided the spark for chasing free-roaming horses. A cowboy probably saw a fine-looking stallion with a bunch of mares. Maybe he spotted the horse the first time purely by accident. The horse captured the cowboy's fancy as something of value. The cowboy probably enlisted a couple of friends to help him run the stallion.

When the horse eluded him, the cowboy was hooked. It was a battle of wits and strategy. The cowboy scouted the country and perhaps put in hours building traps and barriers. The danger of breakneck runs down steep slopes over rocks and brush was exhilarating. Flirting with death was itself addictive. Peter Barnum, a completely professional mustanger, did not take unnecessary chances. He rode a sure-footed mule on chases over rugged country. Even that was no guarantee of safety. Once, after a long wait for the mustang band to reach his position, the mule was so anxious to begin its run that it lost its footing on the first jump and tumbled down the slope with Pete's spur hooked in the saddle.[34] Sagebrush literary figure H. L. Davis had one of his characters fall while running horses in the Wagnerian setting of Walker Lake. He wanted to go out in a blaze of glory, and this was his Valhalla.[35] Cowboys would become so addicted to mustanging that they would risk a valuable horse to run down a scrawny mustang. Riders would jump from exhausted horses and scream insults at animals they had bragged on for years or even shoot the animals because a worthless mustang escaped.

The combination of the lure of the chase, danger, and free horses was a powerful attraction for Native Americans, too. The Indian cowboy as a mustang runner became a fixture on the sagebrush/grasslands. The descriptions by Rufus Steele and Anthony Amaral of mustang running contain numerous references to Indian cowboys.[36] The Paiute from central Nevada who could not mount Captain Simpson's mule had come a long way.

Travelers in the sagebrush/grasslands often come upon a pile of horse manure at natural divides between basins or saddles on long, high ridges. Such dung piles are boundary markers deposited by stallions delineating their territories. Dung piles occur both at high elevations and around springs, though typically they are placed in spots with extensive vistas. They also occur in sheltered resting areas. C. Charles Fisher, who spent thirty years as a range manager on an Indian reservation and became very knowledgeable about feral horses, claims that reports of mustang stal-

lions marking their territory are pure horse manure. He believes all feral horses—stallions, mares, and colts—tend to drop their manure in the
same place; at sites where horses rest during the day, piles develop. According to Charlie, the piles serve a useful purpose. In hard winters, the
horses return and consume the old manure. Horses have a relatively
inefficient digestive system, and their manure contains undigested nutrients.[37]

Territoriality is important in equine ecology and breeding systems
(i.e., the degree to which individual populations or bands inbreed or
outcross). In areas where no visual aids or natural boundaries such as
canyons exist, horses often feed up to half a mile outside their usual
range. When these excursions are made in the presence of neighboring
bands of horses, the question of territory arises. The stud of the group
that is in its own home range runs toward the intruding band. The intruding stud may stand his ground for a short time while his band moves
off, but he usually takes flight behind his mares before the defending
male gets within a hundred yards. The intruder stud pushes his mares
along by putting his head down and weaving it from side to side.[38]

The literature concerning free-roaming horses is full of references
to beautiful stallions outlined against the skyline, but the reality was often scrawny, broken-down mustangs not worth chickenfeed. Which of
these contrasting images is true? Which type of horse actually existed on
the range? The answer is that both are correct; both types of free-roaming horse populations existed.

The sagebrush/grasslands tend to be disjunct environments. Numerous fault block mountain ranges arise from the arid valleys to form islands of favorable environment for large herbivores in a sea of sagebrush. Large herbivore populations tend to be separated into naturally
disjunct groups in this environment. This isolation is a relative thing, for
every old-timer has seen mustangs far out on salt deserts well away from
individual mountains. These sightings are real, but generally the herds
are isolated, and the gene pools of the individual populations are circum-

scribed.[39] When a band became established in a new territory in relative isolation, a process of inbreeding was started. The dominant stallion kept other stallions at bay, heightening the isolation. As his own male offspring reached sexual maturity, he drove them from the band. When the female get of the sire reached maturity, they were bred back by the stallion. Such father-daughter inbreeding resulted in a drastic expression of undesirable characters in the offspring. Horses carry genes for numerous lethal characteristics.[40] Some are congenital and we never see the effect except for the aborted fetus, some are expressed in foals or juveniles, and some are not expressed until the afflicted horse reaches maturity. These lethal characters are recessive characteristics that are carried by heterozygous individuals but are expressed only in the homozygous recessive state. Other, nonlethal, recessive characteristics are expressed in inbreeding populations as well. Strange mustangs have been found on the range with camel-humped backs or startlingly curly hair. There is another effect of inbreeding besides the expression of recessive characters: inbreeding depression. This strange phenomenon is expressed through heritable or genetically controlled characteristics that are highly influenced by the environment. Many of the features of reproduction and early life survival are influenced by inbreeding depression.[41] Inbred mares fail to conceive, or their foals are smaller than normal and lack vigor at birth. The inbred mare lacks mothering instincts and is inattentive to her foal. Inbred populations gradually shrink in body size with successive generations of inbreeding. The actual population size shrinks as well because of poor reproduction, and this shrinkage accentuates the rate of inbreeding. If the rate of inbreeding is not so rapid that the population is lost through expression of lethal genes or infertility, there is a genetic benefit. Only the fittest of the inbred animals survive. The inherently weak are weeded from the population. The harsh environment of the sagebrush/grasslands was a cruel arena for natural selection. The mustangs it produced were small in number and stature, infertile, and incredibly tough.

The inbred bands of free-roaming horses responded to the quality of the environment. If two or more years of above-average precipitation occurred in succession, mustang populations would begin to expand. The expanding populations strained the control of the dominant stallions and naturally led to expansion of the ranges of individual bands. The most dominant stallions captured mares from neighboring populations to add to their harems. The offspring of these captured mares and the stallion were startlingly different; they exhibited heterosis, or hybrid vigor. Hybrid vigor is the reciprocal of inbreeding depression.[42] The greater the depression, the greater the hybrid vigor when two inbred populations are crossed or hybridized. Hybrids are all the things inbred animals are not. Hybrids are larger, more vigorous, and more fertile than their parents. When hybrid mares reach sexual maturity, they have a tremendous influence on the population. The hybrid females are more fertile than inbred mares and produce vigorous offspring. The outbreeding population is a rapidly growing population. Eventually, if the population is left undisturbed, the number of animals will exceed the potential of the environment and the population will crash and lapse back into a cycle of inbreeding. The study of such events as inbreeding and hybrid vigor in actual populations is called population genetics. This branch of science is of vital interest to the managers of natural resources, but it has frightening aspects, too, when one considers that humans constitute an expanding population in a finite environment.[43]

Free-roaming horse populations seldom had the opportunity to pass through cycles of inbreeding and outcrossing without some human interference. Instead of depressing the cycles, however, such interference tended to accentuate them. Humans influenced the breeding systems of free-roaming horses in numerous ways. As mentioned above, there were cycles of predation on free-roaming horse populations in response to general economic conditions. When times were hard, out-of-work cowboys ran mustangs. Increased harvesting decreased the population size. If a predation peak coincided with a population low, the effect on the gene pool was extreme. Humans' most important influence on wild horse populations was through replacing dominant stallions with higher-quality studs. If the free-roaming horse population was on an inbreeding cycle, the sudden introduction of outside genetic material would reverse the trend and produce an outbreeding population. Ranchers in the sagebrush/grassland accentuated this process by importing and releasing a succession of different breeds of horses. Virtually everything from Percherons, Shires, Clydesdales, and Morgans to Shetland ponies could be found on the range at one time or another. Mustang bands ranging on ranches that specialized in raising horses remained on a continuous heterosis high while bands in remote areas might be in the depths of inbreeding depression.

Mustangs became an important item in the Great Basin economy soon after the establishment of ranches. In January 1879, for example, Alvaro Evans received a letter from Ben Payne of Elko requesting help in purchasing two hundred to three hundred mustangs for shipment to Utah.[44]

By 1885 the horse population had risen so high on some portions of the sagebrush/grasslands that the newspapers commented on it. The March 19, 1885, *Reno Evening Gazette* carried an article stating that eastern Oregon was overrun by an estimated 100,000 wild horses. The paper suggested killing the excess animals and using the carcasses for fertilizer.[45] The D. H. McDonald Company proposed building a plant in Elko, Nevada, to boil the carcasses of wild horses for their oil after the hides were salvaged. The Chicago-based firm needed a guarantee of one thousand horses per year at reasonable prices to make the plant feasible.[46]

If horses were a vital part of the ranching environment, horsemen were the symbol of the time. We have already met Pete Barnum, whose exploits as a mustang runner are preserved in the writings of Rufus Steele. Barnum's attitude toward horses is well illustrated in a story related by Anthony Amaral. A ranch foreman failed to return one day, and Barnum was among the party sent to investigate his disappearance.

Tracks indicated the foreman had given lengthy chase to a band of mustangs. Probably frustrated because he could not capture the stallion leading the band, he pulled a revolver and shot the stallion. In the process his own horse tripped and the gun accidentally discharged again, killing the foreman. As the search party sat on their horses around the body a cowboy said, "Poor fellow"; to which Barnum replied, "Poor horse."[47]

On the Sparks-Tinnin ranches in far northeastern Nevada, a horseman of special stature came along during the 1880s. Henry Harris was born in Williamson County, Texas, and came to Nevada to work for John Sparks in 1885 as a seventeen-year-old houseboy. He did not remain a houseboy very long, for he had natural talent as a rider and roper. In later years the byword of Elko County cowboys was, "If it had hair, Henry could ride it."[48] One other thing: Henry Harris was black. In fact, John Sparks employed many black cowboys on his ranches in Texas and Nevada. Will Pickett, the original bulldogger of wild west shows, got his start on Sparks's Texas ranch.[49] Henry Harris became foreman of a ranch crew of black cowboys on the Boars' Nest Ranch on Salmon Falls Creek and later became one of two wagon bosses for Sparks-Harrell. He ran the roundup wagon that worked the Idaho side of the range, and Tap Duncan ran another on the Nevada side.[50]

On one occasion Henry had the Sparks-Harrell wagon on Devil Creek for the spring roundup. He was in his bedroll when the horse wrangler mounted to bring in the horse herd for the crew. After the wrangler's horse threw him twice, Henry jumped from his bedroll, pulled on his boots, and rode the horse to a standstill in his red underwear. The boots were necessary for the big silver spurs he used to rake the horse from flank to shoulder. Henry dismounted, none too gently set the wrangler on the horse, and climbed back into his bedroll, still wearing his boots and spurs.[51]

Henry had an incredibly casual relationship with horses. He probably had more runaways and horse accidents than any other cowboy in the sagebrush/grasslands. He would hook a team of mustangs to a wagon or mower, put them through maneuvers as if they were a well-broke team, and then deliberately push the team to the edge of a runaway. A big sorrel horse called Ben roamed the Snake River valley slopes of the Sparks-Harrell range. Ben could be caught without too much trouble, and it became popular sport for farm boys from the growing irrigated areas along the Snake to try and ride him. Ben's reputation grew with each farm boy he piled. The ranch crew eventually brought Ben into headquarters, where he was happy to munch wild hay from a manger under a lean-to beside the horse corral. Naturally, there were sly remarks about the possibility of Henry taking a ride on the best bucking horse in the country. Henry waited until he had a good crowd of onlookers before he slipped into the corral, took a mane-hold on Ben, and threw himself on bareback under the lean-to! Everyone stopped breathing in expectation of Henry being bucked through the shed roof, but old Ben just kept munching hay. Perhaps Henry's audacity was just too much for him.[52]

Henry was a master with a rawhide lariat. One day, he and Foley, another foreman, were roping and throwing young horses entering a corral at San Jacinto. As each horse came through the gate, they took turns catching its two front legs and throwing it. The cowboys were awed by this display of roping expertise. Horse after horse was crowded through the gate, but neither roper missed. Finally, maybe in response to the awe of their audience, the two ropers got out of synchrony and caught the same horse. The horse flipped and broke his neck, leaving two sheepish foremen with slack ropes and a laughing crew.[53]

After the hard winter of 1889–90, many of the black Texas cowboys drifted away. Henry stayed and remained a noted figure in the ranching industry until he died in 1937.[54] He was not completely alone, for his brother, Leige, also came to Nevada from Texas. There were others, too. Adelaide Hawes refers to occasional black cowboys in *The Valley of Tall Grass*. Henry Harris explained to the Harrell boys one reason why most of the black cowboys left. They asked with the curiosity of youth why Henry never married. He laughed and said, "There sure are a lot of black gals hanging around the sagebrush."[55] Leige solved this problem by marrying Lizard, one of the daughters of Indian Mike, a Bannock Indian who never went to the Fort Hall reservation. The family lived largely on Sparks-Tinnin and later Sparks-Harrell rangeland from Rock Creek at the Snake to the upper ranches of Salmon Falls Creek, eking out an existence as hunter/gatherers, mustang runners, and seasonal hay hands on Sparks-Harrell ranches. The marriage of Lizard to the tall, handsome Leige was an astute political move because of Henry's position with the ranch management. It ensured permission to roam on the company's rangeland and occasional credit at the company store at San Jacinto. Ethnologically it became a tangled web, especially when Indian Mike's family fell on hard times and was almost wiped out in the last Indian uprising of the West just before World War I.[56]

When he was an old man, Henry was riding a big sorrel horse in the Cottonwood field at San Jacinto with a crew on working horses. The sorrel suddenly bucked and Henry hit the ground hard. The cowboys were shocked beyond words. The impossible had happened; Henry Harris had been bucked from a horse. The shock was still apparent in Andy Harrell's face when he recounted the incident to me decades after it happened. Someone caught the sorrel and Henry remounted without a word. The incident verified the old maxim that "never was there a horse that could not be rode, never a cowboy that could not be throwed," but the price of the proof was too high.

It seems surprising that there are no records, published or personal, indicating that John Sparks was a fancier of horses. He was a decided connoisseur of exotic animals and a noted breeder of dogs, including foxhounds and greyhound-wolfhound crosses for running coyotes.[57] He loved to hot-rod around in his custom-made buggy. Despite his obvious love for well-bred cattle and the finer things in life, however, John Sparks apparently had no interest in breeding horses. Sparks had at least one bad experience while riding horseback. During a Fourth of July parade in Tonopah, Nevada, his horse suddenly shied and he fell from the saddle to the street. He landed in an undignified position on the tails of his frocked coat, but his tall silk hat remained in place.[58]

Banker Jackson Graves once got a glimpse of Sparks's attitude toward mustangs. Approaching the H-D Ranch, they saw a band of some thirty mustangs along the fence edging the hay fields. When they reached the ranch headquarters, Sparks immediately ordered the crew to get rifles and kill the horses.[59]

Many ranchers came to share Sparks's attitude. The ranges had been degraded by two decades of excessive grazing, and ranchers were feeling the pressure of competition from sheep operations. There was no room on the range for freeloading, free-roaming mustangs. During the white winter of 1889–90, the ranchers in Long Valley along the California-Nevada border tried to take advantage of the exceptional snow cover to rid Fort Sage Mountain of some four hundred wild horses. Even though the snow forced the horses down off the slopes, the ranchers were not successful in getting rid of all the pesky animals.[60]

During the 1890s wild horses were hunted for hog feed and for sale to canneries. In 1893 Senator Charles Kaiser of Churchill County, Nevada, proposed a bill permitting the uncontrolled destruction of wild horses. Ranchers paid twenty-five cents per carcass and supplied rifle ammunition to the hunters.[61] Buyers for eastern canneries were eager to get the meat; some even visited such remote areas as Lincoln County, Nevada, looking for horses. One Maine company built a canning plant near Mud Lake in Humboldt County, Nevada. Its chief supply of horsemeat came from the Pyramid Lake Indian Reservation.[62]

Despite being shot for hog food and canned for dog food, wild horses still had value at the end of the century. Lieutenant Colonel Marshall of the U.S. Army, for example, was buying remount horses in Nevada in 1899 to add to the two carloads he had purchased in northeastern California.[63]

* * *

At the turn of the century, many people in the Intermountain area were proclaiming the end of the free-roaming horse. For a brief time horses had reentered the geographical area in which their ancestors had evolved and had survived the harsh land on its own terms. We now know that the prophets proclaiming the demise of free-roaming horses were premature. They vastly underestimated the mustang's ability to adapt to the sagebrush environment.

The pioneer horsemen of the sagebrush/grasslands have all reached the end of their tenure. Henry Harris is buried in Twin Falls, Idaho, beneath a simple stone marked "Pioneer Cowboy." Every spring the great-nephews of Jasper Harrell, Newton and Andy Harrell, ramrod straight from a life in the saddle, go out into the sagebrush and pick red Indian paintbrush, yellow balsamroot, and royal blue larkspur flowers to put on Henry Harris's grave, memorial to the famed black horseman of a silver and gray environment.

Part IV **The Land Answers**

The turn of the century brought sweeping changes for both land and man in the Great Basin. The cold desert's sagebrush/grasslands had only a limited potential to support life, and it was often exceeded in the grand experiment to establish ranching. This fragile ecosystem did not bend to accommodate man and his herds; it shattered. The demand for silver depleted the Comstock Lode, and just as surely the demand for rangelands destroyed the sagebrush/grasslands.

CHAPTER 13

The Passing of the Old Guard

The 1890s were strange years in Nevada. Silver mining was in a terrible slump, and the decade began with a major depression. The economy and population of Nevada declined until the question of revoking statehood was considered. The winter of 1889–90 had crippled the livestock industry and altered the methods of raising livestock. The bloom was definitely off the sagebrush/grasslands.

After the disastrous winter of 1889–90, John Tinnin left the Sparks-Tinnin Company to ranch in Nebraska, and the company was restructured and incorporated as Sparks-Harrell. Jasper Harrell was the major owner of the new corporation in partnership with John Sparks. Possibly, Sparks and Tinnin never finished paying for the ranches they had purchased from Jasper Harrell in 1881. The sale had called for mortgage payments of $100,000 annually until the $900,000 purchase price plus interest was paid. Even if the payments had been made on time, Sparks-Tinnin would still have had payments running on the original mortgage. In a sense, Jasper Harrell reassumed control of the vast Nevada and Idaho holdings he had founded.

Harrell continued a vigorous program of water development to bring more land under cultivation on Sparks-Harrell ranches. As time went by, John Sparks drifted away from his ranches in northeastern Nevada and Idaho to concentrate on the Alamo Stock Farm at Reno. Jasper's son, Andrew J. Harrell, took an increasingly active role in management of the Sparks-Harrell ranches during the 1890s.

A resolution adopted by the Sparks-Harrell Company at the annual meeting in Visalia, California, on February 7, 1895, illustrates the desperate economic situation in Nevada: "That all persons employed by the superintendent, any foreman, or any employer having authority to engage the services of any person in behalf of this Company, on or after the 15th day of February 1895; that the salary of said new employee, or person hired, shall not exceed the sum of thirty ($30.00) dollars per month."[1]

Despite his interest in the Alamo Stock Farm, John Sparks was still president of the Sparks-Harrell Company midway through the 1890s. At that time, and in subsequent accounts, Sparks was viewed as the company's driving and managing force.[2] That may not have been the case. A letter written by A. J. Harrell in 1896 to his cousin Louis Harrell sheds some light on the operations.

I am disappointed that Harris and Duncan did not gather more than 1,100 head of cattle, as I thought there were a great many more than that number there; when you get this, write me how many cattle you think are left on that range, and if the boys said they got a good clean gathering. And you go to the desert as we talked it over, and keep a good watch on the cattle we have left there, and organize an outfit big enough to make a clean gathering next spring; do all this without any further orders, for we may not have a chance to instruct you further before work begins. I have not heard from Sparks since I saw you, but think you will see him by the time you get this, and when you do, talk over with him what I have instructed you to do, and if he disapproves of it, do as he says; but if you do not see him, carry out my instructions until you get further orders.

This letter suggests that John Sparks was still a power in Sparks-Harrell but was not in close enough touch with ranch operations to plan and direct such important events as roundups.[3]

Some basic environmental constraints began to pressure the company. The winter of 1889–90 made it obvious that hay reserves were necessary to guard against excessive winter losses. Raising hay is closely tied to irrigation. Although the white winter should have strengthened the position of landowners who controlled irrigation water, another livestock production pattern that avoided the need for hay production was becoming established: raising sheep. Sheep are more efficient browsers, require less water per equivalent animal unit, and can utilize winter range while depending on snow for water. Also, sheep raising required less capital. If hay production requirements for cattle were included, it was ten times as costly to go into the cattle business as the sheep business. Cull ewes had less residual value than cull brood cows, but their initial cost was proportionately less. Further, many range sheep operators were completely nomadic. They owned no land and had none of the expenses associated with landownership.[4]

Cattle ranchers faced competition not only on winter ranges, but also on higher-elevation summer ranges. Summer range became the most limiting factor in livestock production. Not only were cattle and sheep degrading the same overgrazed range, they were destroying the vegetative cover of the watersheds that provided the runoff that supported agriculture and hay production.

Sparks-Harrell faced continued pressure from the northeast. The spread of range sheep appeared to be a threat to "their" rangelands. More than 90 percent of the company's rangeland was public land, and the major partners were absentee landlords for the land they did own. The mountain corridor across southeastern Idaho that provided a route for the California Trail also provided the environment for the spread of relatively small sheep operations. Many of those running sheep were members of tightly knit Mormon communities.

Jasper Harrell had virtually founded the ranching industry in north-eastern Nevada and south-central Idaho. If he could have obtained title to all the land where he ran livestock, he doubtless would have done so. But his and his partner's possessory claim to the range lacked legal status. Sparks-Harrell's principal hold on the rangelands was through force and prior arrival.

Sparks-Harrell drew a line across the Goose Creek Basin and let sheep men know that their herds would not be permitted west or south of this "deadline." By mid-1895 the sheep men had become more aggressive and took their flocks across the dividing line, inviting consequences. Sparks-Harrell hired "outside men" to deal with them. Among these gunmen were Jack Davis, Billy Majors, Fred Gleason, and William Majors.[5] Their instructions, passed on through James E. Bower, the superintendent, and Joe Langford, the range foreman, were: "Keep the sheep back. Don't kill, but shoot to wound, if necessary. Use what measures you think are best. If you do have to kill, the company will stand behind you. There is plenty of money and backing, and the company won't desert you regardless of what happens." Viewed in the context of the 1890s, such instructions were not unusual. Captains of industry resorted to similar tactics in fighting the labor movement.

John Sparks was no stranger to "frontier law." As a Texas Ranger, Sparks had fought Comanche Indians while defending early settlers' homes and farms. In 1875, on Wyoming's North Platte River, he had established a ranch on the frontier and defended it against the Sioux. When the Cheyenne and Sioux kicked up trouble that same year, Wyoming Territory Governor Thayer gave Sparks a captain's commission with guns and ammunition to protect the settlers.[6] The Wyoming Stock Growers Association, to which Sparks belonged, had employed detectives to help cope with rustling, eventually leading to the Johnson County Range War. In Nevada, Sparks employed T. M. Overfelt, previously employed by the Wyoming Stock Growers Association as a range detective. Overfelt was killed near Elko by a runaway team under rather suspicious circumstances.

Jack Davis was a central character in the drama that subsequently unfolded on Deadline Ridge. Davis was an unlikely looking gunman. Short and slight of build with a ruddy face and sandy mustache, Jack Davis was likable, well mannered, and kind. He was also a complex personality who compulsively talked himself into trouble. He bragged endlessly about his prowess as a gunman, which he called "cutting it in smoke." Despite his willingness to talk about himself, little definite was known of his background. If he could be believed, he had been born in four different states, fought Apaches in Arizona and revolutions in South America, and worked with Cecil Rhodes in South Africa. Also, he had been a miner in the early 1890s working in the Silver City district of Owyhee County, Idaho. Rumors of a diamond strike in some nearby hills had prompted Davis to go and seek his fortune as a prospector. The diamond fields proved an illusion, but Jack seldom tired of telling what he would have done if he had found the diamonds. A cowboy friend christened him "Diamondfield," and the name stuck.

When Diamondfield Jack started patrolling range for Sparks-Harrell, he spared no words telling isolated herders how dangerous he was and what he would do if they crossed the deadline. His pattern was to ask the herders who they worked for. Regardless of what they answered, he told them if they worked for a different outfit he would have had to shoot them. The sheep men soon discovered that Jack was always gunning for a sheep man who was somewhere else at the moment. Often, in the process of threatening some poor herder, Diamondfield Jack would break off the argument and have supper with his adversary if the grub smelled good.

On one occasion Bill Tolman, a prominent Idaho sheep man, boldly rode up to a Shoshone Basin Sparks-Harrell line shack and confronted Jack with a rifle. Diamondfield engaged Tolman in a long argument, after which he drew his .45 and shot Tolman through the shoulder. Jack then treated the fallen sheep man's wound and made him as comfortable as possible. He later transported Tolman to a sheep camp so he could be moved to Oakley, Idaho, for treatment. The sight of the wounded sheep man being carried home was enough to discourage other herders from crossing the deadline.

Apparently, Diamondfield Jack realized he was in trouble, because he collected his pay and headed for Wells, Nevada. In the warmth of Fisher's saloon and Alice Wood's palace of pleasure he was soon happily telling everyone he had been "up in Idaho shooting sheepherders." On February 16, 1896, a sheepherder on Deep Creek noticed scattered bands of sheep. When he went to investigate, he encountered a grisly scene: two young Mormon sheepherders, John Wilson and Daniel Cummings, had been shot to death in their wagon. Not much imagination was required on the part of Cassia County sheep men to make Diamondfield Jack the prime suspect.

A century after the killings, the two sensational trials that followed, the pardon board hearings, and numerous newspaper articles and books, the question of who killed Wilson and Cummings is still a volatile issue in the Intermountain area. According to their own confessions and the available evidence, James E. Bower and Jeff Gray killed the men in self-defense. Bower was the general manager of the Sparks-Harrell Company. He was also the cowboy who had first glimpsed the Snake River valley grasslands and hurried back to Jasper Harrell with news that helped found the Harrell ranching empire.[7] Bower had helped build the first schoolhouse in the area and organized the first Sunday school classes. He was also an independent cattleman in Idaho. Jeff Gray was a local boy who had grown up with the area sheepherders.

Bower and Gray signed affidavits swearing that they had stopped at a sheep camp on the morning of February 4, 1896, to ask the two herders if they intended to move their sheep onto Sparks-Harrell range. As all four sat around the stove in the sheep wagon, hot words were exchanged and the men grappled. When it was over, the herders had been shot. Thinking that only one man was wounded and that herders in neighboring sheep camps would have heard the shots and be in hot pursuit, Bower and Gray rode away. They told several cattlemen about the incident before and after they learned that both men had been killed. Gray confided to A. D. Norton, the pioneer livestock man of Rock Creek, Idaho. Bower presumably told John Sparks and Jasper Harrell.

But their affidavits did not come out until October 13, 1898—after Diamondfield Jack had twice been tried and sentenced to hang! Both times, he waited under the gallows before a rider galloped in with a reprieve from the governor of Idaho. The trials featured top defense lawyers paid for by Sparks-Harrell and special prosecutors rumored to be sponsored by the Mormon Church. The trials and resulting publicity helped establish the political career of William E. Borah, the "Lion of Idaho," as a U.S. senator.

Diamondfield Jack based his defense on the timing of his physical presence at various ranches on Salmon Falls Creek. Many witnesses testified that they had seen him at the Brown Ranch in Idaho and at the Sparks-Harrell Boar's Nest Ranch on the Nevada side of the state line around the time of the murders. But the question became, "Did Diamondfield Jack ride at a normal speed from ranch to ranch or did he dash to Deep Creek, kill the sheepmen, and continue to the Boar's Nest Ranch?"[8]

Diamondfield Jack's trials were held in Cassia County, Idaho, with juries composed of sheep men. Conviction was expected. The second trial was followed by a long period of appeals, stays of execution, and meetings with the Idaho Board of Pardons. After Bower's and Gray's confessions were made public, it was obvious that Diamondfield Jack was being held for the wrong crime. On December 17, 1902, the Idaho Board of Pardons granted Diamondfield Jack a full pardon. He had spent six years in prison for a crime he apparently did not commit. He had very nearly talked himself to death.

Many rumors came out of Diamondfield Jack's case. One was that the Mormon Church helped finance his prosecution. After Jack's pardon, there were persistent rumors that the pardon had cost John Sparks a small fortune. It was no secret that the company had financed the cost of Diamondfield's defense and appeals. If bribery of the pardon board was involved, however, it has escaped discovery despite persistent investigations by historians. Apparently, Diamondfield Jack was paid well by Sparks-Harrell for his six years in prison. After being released from the Idaho State Prison in Boise, he reappeared in south-central Nevada during the Goldfield and Tonopah booms in the early twentieth century and made money in mining and land speculation. He also proved useful to the Wingfield-Sparks interest in fighting labor unrest in the mines and mills. He developed the mining camp of Diamondfield, Nevada. When the boom in central Nevada mining died, he became a successful real estate developer in southern California. He died in 1949 after being hit by a taxi while on a gambling trip to Las Vegas.

The case of Diamondfield Jack is symptomatic of the problems the livestock industry was having at the turn of the century. Basic forage resources had been severely depleted by years of overgrazing. This, coupled with the severe winter of 1889–90, had nearly destroyed the industry and forced a redesign of forage utilization centered on hay production.

Established ranchers were caught in a paradox. The taxes on their land were greater than their potential income under state and national economic conditions.[9] Those who could not afford to buy the rangeland necessary to support their operations—and that was nearly everyone—had to protect their possessory rights on the public rangeland. Sparks and Harrell were protecting these rights when they hired outside men such as Jack Davis.

The same desire expressed through political action led to pressure from the large-scale ranchers to establish national forests on the higher mountain ranges in the early twentieth century. Grazing permits on such lands were allotted on the basis of history of use and ownership of commensurate property to support the livestock when not on the national forest. Both requirements favored established ranchers. Essentially, the federal government would provide de facto recognition of the possessory interest of the ranchers. But the national forests were at least a decade away in the 1890s.

The rise of the range sheep industry was possible because the cattle ranchers lacked legal avenues and capital to acquire title to the extensive acreage essential to cattle operations in the sagebrush/grasslands. There was one obvious solution for Sparks-Harrell to meet the competition from range sheep. The company could have gone into the sheep business. The Utah Construction Company did just that for the first half of the twentieth century on the former Sparks-Harrell ranches.[10] Why did Sparks-Harrell ignore that option? Lack of knowledge about the sheep business may have contributed, but traditional prejudice against sheep and sheepherders was probably a bigger factor.

The depth of the ill feelings that existed between established cattle ranchers and sheep operators is apparent in the Diamondfield Jack case and from other sources as well. And the hostility was found in high places. In his address to the state in 1896, for example, Governor Reinhold Sadler stated: "With the exception of a comparatively few million acres, the entire State of Nevada must remain a stock raising state and the greatest benefit which she could not [sic] receive at the hands of Congress would be some manner of legislation looking to the preservation of her ranges to prevent them from being laid waste by foreigners—men who are not citizens, never intend to become such, and who use our state only to ruin it, and filch from our people their national heritage."[11]

Foreigners were very much a part of the controversy. Several large local cattle operations branched into the sheep business with scarcely a comment from their neighbors. It was the new sheep outfits, often owned by Basque herders, that especially raised the anger of cattle ranchers. Many of these sheep outfits were completely migratory and owned no base property at all in Nevada. Even government publications such as Griffith's survey of the northern Great Basin at the end of the 1890s slurred the religion and national origin of the migratory sheepherders.[12]

John Sparks was always ready to use his reputation and standing to advance the company's interests. He used the wildlife resources of the Sparks-Harrell rangelands to entertain and curry favor with a variety of powerful bankers, judges, and financiers.[13] In late August, guest hunters from Cheyenne, Salt Lake, Los Angeles, and San Francisco would drop off the train in Wells, Nevada, to be ushered to the Sparks-Harrell ranches. The hunters traveled in style, for Sparks-Harrell provided a cook, camp tenders, a horse wrangler, and forty extra riding horses.[14] The big game hunters were after mule deer. Bucks had antlers still in velvet in late August, and it is difficult to see how the hunters kept meat from spoiling in the hot weather typical of early autumn. There were abundant sage grouse to shoot around the stringer meadows, and the more venturesome hunters tried for pronghorns in the valleys.

Always the extrovert, John Sparks carried his special rifle, "Alcade," on these trips and displayed notches in the stock that represented deer, bear, elk, and buffalo he had shot. Alcade was a Sharp .50 rifle, or "buffalo gun." Once he demonstrated how well he could shoot the old rifle by knocking down a buck at a reputed six hundred yards.[15] One gets the impression that Sparks enjoyed playing the role of king of the sagebrush ranchers before his audience of powerful and influential friends. He often served them a special Texas breakfast consisting of a baked bull's head split so the guests could scoop out the brains.

The visiting hunters generally had good luck with sage grouse, often killing far more than they could consume. The sage grouse is native to the sagebrush/grasslands, and apparently there were enough meadows left in that environment in the 1890s to support sage grouse chicks. Mule deer were not abundant then, even in the high mountains of extreme northeastern Nevada. On some years, the entire party was lucky to get a single shot at a buck. Shrubs that provided forage for browsing deer were increasing on the sagebrush/grasslands, especially in the aftermath of the winter of 1889–90, and mule deer populations were responding to this increase, but populations had not yet increased to the point where deer were abundant. On one trip Sparks blamed the lack of deer on the presence of Indians on his range.[16]

Intermountain ranges during the 1890s were also affected by the recurring drought in California and the movement of cattle across the Sierra Nevada. California experienced varying degrees of drought from 1895 to 1900. In June 1895 starving cattle were arriving in Reno from California, and by midsummer John Sparks was reporting that forage was extremely short on his Nevada rangelands.[17] The California ranges were again dry in 1896. The problems of California stockmen were compounded by the outbreak of Texas fever in southern California. Secretary of Agriculture J. Sterling Morton clamped a federal quarantine on cattle shipments out of Nevada. Nevada stockmen hailed the quarantine as a good thing. Everyone was interested in protecting the health of his stock, of course, but the quarantine also happened to keep the giant firm of Miller and Lux from shipping thousands of cattle from California to Nevada, where they traditionally summered.[18] The federal government lifted the quarantine in late winter, and Nevada cattlemen immediately pressured Governor Sadler to declare a state quarantine against California cattle. The governor, a cattleman from Eureka County, Nevada, responded favorably to the request and immediately found himself in hot water. J. H. Budd, the governor of California, under similar pressure, personally applied to Sadler to lift the quarantine to save starving cattle. Miller and Lux was a powerful firm in California with great political influence. Virtually every cattleman in Nevada signed petitions to keep the quarantine in effect. Surprisingly, John Sparks was quiet about the controversy. Sparks knew all about Texas fever, as his herd had carried it from Texas to Virginia in the 1860s. As the quarantine dragged on to the turn of the century, Sparks was criticized for importing uninspected purebred Jersey cows from Texas to his Alamo Stock Farm.

The quarantine controversy heated up in 1899 when Miller and Lux requested permission to ship ten thousand head of cattle through western Nevada to stock a ranch in southeastern Oregon that the company had purchased from the Sharon Estate. Miller and Lux reported that the cattle had been gathered off the California winter range in such poor condition that they did not have the strength to walk to their new range. The company wanted to ship the cattle to Reno on the Southern Pacific and reload them in narrow-gauge railcars for shipment north on the Nevada, California and Oregon Railroad. Nevada newspapers jumped on Governor Sadler even before he replied to Miller and Lux. Essentially, the quarantine became a method for established ranchers in Nevada to extend their de facto control over the public rangelands. The ranchers even attempted to extend the quarantine to animals that were not susceptible to Texas fever, especially California sheep. In 1899 the governor of Idaho wrote Governor Sadler requesting information on the quarantine. He proposed to extend it to Idaho as suggested by that prominent cattleman John Sparks, who happened to be visiting his office.[19]

John Sparks remained a national figure and embellished his reputation as a spokesman for the livestock industry.[20] He served on the public lands committee that lobbied for a federal policy allowing ranchers to lease public lands. Sparks sponsored Dr. J. E. Stubbs, president of the University of Nevada, as a delegate to the 1898 National Stock Growers Convention. The Agricultural Experiment Station of the University of Nevada had been founded in 1887 under the provisions of the Hatch Act.[21] The concept of universities conducting agricultural research was conceived during this period, and Stubbs went to the convention to convince stockmen of its value.

Among the other delegates at the 1898 National Stock Growers Convention was the familiar name of John Tinnin, listed as a delegate from South Dakota but residing at Gordon, Nebraska. The Nevada delegates read like a who's who of the nineteenth-century livestock industry: Col. Hardesty, J. R. Bradley, N. H. A. Mason, J. J. Altube, and A. C. Cleveland. By 1900 the changes on the range were reflected in the delegate structure of the National Stock Growers Association: sheep men were elected to serve as delegates to the national meeting. W. H. Poulton of Oakley, Cassia County, Idaho, was a delegate from the state Sheep and Wool Growers Association.[22] This organization may have helped heal some of the wounds created by the Diamondfield Jack affair.

Among the presentations made by delegates at the 1898 meeting of the National Stock Growers Association was one by William Byers, an obscure rancher who spoke from the heart and touched the soul of a range livestock industry still reeling from the white winter of 1889–90. Speaking under the title "The Humane Treatment of the Range Stock," Byers first stressed the value of conserving forage for the use of wintering cattle. He pointed out that one-half ton of hay made a yearling or two-year-old worth five to eight dollars more in the spring than a calf wintered on the range without supplemental feeding. He suggested that the cost of hay production on the ranch should not be more than two to three dollars per ton. Byers concluded with a plea for humane treatment of livestock: *"Whenever animals are under man's control, it is his duty to see that they do not suffer from any cause which he is able to remove."*[23]

The operational style of Sparks-Harrell began to change during the mid-1890s. After 1896 the Sparks-Harrell roundup wagon no longer operated on the Owyhee Desert. This vast area, which contained much winter range, had been a stronghold of Jasper Harrell's operations in the 1870s.[24] Competition from smaller ranchers in Idaho, especially in the Bruneau Valley, who had expanded their operations after the winter of 1889–90 forced Sparks and Harrell to downsize their operation. While John Sparks was increasing his political activity and interest in the Alamo Stock Farm, Andrew J. Harrell was enlarging the irrigation ditch network on the company's land and bringing additional land into hay production.

Some accounts of the Sparks-Harrell operation describe A. J. Harrell as a Bay Area butcher and a California playboy whom John Sparks tolerated while he guided the fortunes of the company.[25] In actuality, A. J. Harrell, who had a business college education and experience gained in ranching, real estate, and banking, was the driving force behind Sparks-Harrell during the 1890s. Behind A. J. was the flint-hard figure of old Jasper Harrell. In 1891 Jasper and A. J. sold the Harrell & Son Bank of Visalia to the Producers Bank and A. J. devoted all his attention to the firm's Nevada interests.[26] In 1899 A. J. Harrell moved his wife and two children from Visalia to Palo Alto so he could be closer to San Francisco, the financial and cultural center of California.

Harrell's influence on the Sparks-Harrell operation was apparent when California banker Jackson Graves visited the ranches in the mid-1890s. A. J.'s improvements were immediately obvious. The H-D Ranch had eighteen hundred acres of irrigated meadow and alfalfa. The Hubbard and Vineyard Ranches both had substantial pole buildings with dirt roofs. At the Hubbard, three thousand acres of sagebrush had been converted to irrigated fields. Graves saw immense stacks of hay at all the ranches. Some of the stacks had been carried over from previous seasons as insurance against another winter like 1889–90.[27]

The year 1901 marked the end of the introduction and expansion period for cattle in the sagebrush/grasslands. John Sparks sold his interest in the Sparks-Harrell ranches to A. J. Harrell. The April 4, 1901, issue of the *Reno Evening Gazette,* headlined "The Sparks Sale—Over a Million Paid for His Cattle and Ranch Property," indicated that the deal had been concluded in Salt Lake City five days prior to the announcement. John Sparks had sold to A. J. Harrell of Visalia, California, 20,000 head of range cattle along with half interest in 700,000 acres of land and a lease on 700,000 additional acres. Sparks refused to discuss the purchase price with reporters, but it was rumored that he received $500,000 in cash and 12,000 acres of Texas cotton land for his half interest in Sparks-Harrell. The Salt Lake papers quoted Sparks as saying, "I plan to devote some attention to raising cotton on my little 12,000 acre Texas ranch. . . . Cotton is King you know, and if I raise enough of it, I may make some money—there is no telling."[28] The firm continued until 1908 as Sparks-Harrell.[29] In a widely quoted interview published in *Harper's Weekly* in 1902, Sparks strongly implied that he was still the owner of the Sparks-Harrell ranches.[30]

Although he told his interviewer that he had yielded to a tempting offer in giving up Sparks-Harrell, the sale may not have been by choice. Considering his later financial problems, John Sparks probably was unable to pay his mortgage debts to the Harrells and lost his ranching empire by default.[31]

On May 13, 1901, Jasper Harrell died. A. J. Harrell inherited sole ownership of 175,000 acres in the central Great Basin and 30,000 cattle that ranged on 3 million acres.[32] At its high point in the 1880s, this ranching empire had, by some reports, 150,000 head of cattle. This is one of the most striking statistics showing how overstocked the ranges were in the late 1800s. In 1901 brood cows were getting one-half their forage requirement from irrigated lands in the form of hay and crop aftermath. In the 1880s a reported five times as many cows were dependent on the range.

While President Stubbs was busy selling university research at the National Stock Growers Association, there were other moves under way to enhance the western rangelands as well. Secretary of Agriculture Wilson announced in 1898 that the USDA had sent Professor Niels Hansen from Brookings, South Dakota, to Central Asia to collect plants for revegetating semiarid areas in the West.[33]

Explorations were also being carried out closer to home. The USDA sponsored a survey of the western range to try to determine the nature of the resource. David Griffith, G. Vadey, F. Lamson-Scribner, A. Nelson, and G. Smith for the USDA; F. H. Hillmand and P. B. Kennedy of the University of Nevada; and H. T. French of the University of Idaho began to describe the nature and extent of the grazing resource.

In 1904 A. J. Harrell was experimenting with grasses from the Russian steppe on his properties in Nevada in an attempt to fill the niche left open by the destruction of the native perennial grasses.[34] The grass he was using was described as a "recent import from Russia." Although it is unlikely at that early date, the grass may have been crested wheatgrass introduced by Professor Hansen. By the middle of the twentieth century, millions of acres of severely degraded sagebrush rangelands had been seeded to this species.

Native shrubs, especially sagebrush, had partly preempted the environmental potential released by the destruction of the perennial grasses. The sagebrush communities became stark, shrub-dominated landscapes without sufficient understories to support anything but marginal livestock production. Most important, the sagebrush-dominated communities were extremely stable for the shrubs because there were not enough herbaceous understory plants to carry fire through the shrub communities. The shrubs, however, did not have the ecological amplitude to fill the potential of the communities; some resources remained unused. But biological near-vacuums do not last.

One development was already in process. In the Red Desert of Wyoming a new plant species had been discovered—a spiny, coarse herb that uprooted when it was mature and tumbled across the landscape spreading seeds. Known as Russian thistle, it had been accidentally introduced to South Dakota in the 1870s and was to become the first of the alien annual weeds to invade the sagebrush/grasslands.[35]

With the death of Jasper Harrell soon after the turn of the century, the days of the empire he founded were numbered. The remaining principal characters did not last the decade. John Sparks had a spectacular rise in Nevada politics and was twice elected governor after an unsuccessful race for the Senate. But the colorful promoter was himself caught in a fraudulent mining promotion and lost a great deal of money—both his own and money belonging to friends.[36] The last public acts in his action-filled life occurred during the strike and labor controversy between Tonopah and Goldfield mine owners and the miners' union, when Sparks got federal troops from President Roosevelt to control the area. He arose from his sickbed in midwinter and rode sixty-five miles to do what he could on the scene, but was rebuffed by Teddy Roosevelt, who called back the troops. Broke and broken, Honest John Sparks went home to the

Alamo Stock Farm and died in 1908 of Bright's disease—and, some said, a broken heart. The year also saw the passing of A. J. Harrell.

Banker Jackson Graves mourned the loss of both friends. His last view of A. J. came while he was visiting an estate at Lake Tahoe, where it was becoming quite fashionable to have a summer home, and went down to the dock to see the steamer *Talac,* the pride of the lake. As the boat pulled away in the mist, he saw a forlorn figure wrapped in a heavy coat standing on the deck whom he recognized as A. J. Harrell. When Graves returned to Los Angeles, he was shocked to hear that Harrell had passed away. Later that year Jackson Graves visited Reno and drove south of town to pay his respects to John Sparks's widow. He found the front door of the beautiful Sparks home boarded shut and the registered Herefords and exotic animals gone. John Sparks had died virtually bankrupt. The Sparks-Harrell ranches passed through intermediary ownership before being sold as a block to the Utah Construction Company, under whose ownership they remained until after World War II.

Writing a decade after John Sparks's death, John Clay listed six reasons for his financial failure: (1) low stock prices, (2) winter losses, (3) distance from market, (4) politics, (5) mining, and (6) purebred cattle.[37] John Sparks had been the dominant figure in livestock in the Great Basin during the last quarter of the nineteenth century. Two years after his death, his beautiful home at the Alamo Stock Farm was sold to William Moffat, the dominant figure in the livestock industry for the first half of the twentieth century.

The pioneers of the livestock business passed away one by one. A few characters in the drama, such as Indian Mike and Henry Harris, overlapped into the twentieth century.

The pristine vegetation of the sagebrush/grasslands was gone. It was still possible then to return the plant communities to their pristine composition because the spread of alien weeds had not yet occurred. The range was in trouble, but no one could correct the problem.

*   *   *

There are now vast areas of sagebrush/grasslands in relatively good condition that could greatly benefit from grazing management. At the end of the nineteenth century, there were areas at least as abundant as those of today. What if the federal government had sold or leased the sagebrush rangelands to established ranchers? Probably the results would have been mixed, but two such areas offer facts for speculation. One is a huge block of land on Gollaher Mountain that Sparks-Harrell obtained from state select land. The other, north of Battle Mountain, Nevada, is known as St. John's field.[38] Both areas have been in private ownership since the nineteenth century and consist of large blocks of upland range. Both are examples of near-pristine sagebrush/grasslands. This does not mean to imply that private ownership would have resulted in better range conditions across the landscape, but it is food for thought.

Expansion of the cattle industry into the sagebrush/grasslands was a grand experiment in which herdsmen boldly ventured into an environment then considered beyond the potential of agricultural enterprise. One result was the birth of a system in which cattle graze on extensive rangelands during the spring and summer and eat hay or graze crop aftermath in the fall and winter. Hay is essential for this system to operate, but it is produced on only a fraction of the irrigable landscape. The second result of the experiment was the destruction—within a mere forty years—of the sagebrush/grasslands vegetation born in the wild climatic fluctuations of the Pleistocene and scantily nurtured by the post–Ice Age aridity of the Intermountain area.

Two basic factors contributed to the destruction of the sagebrush/grasslands. First, the landscape-dominating shrub known as big sagebrush was protected from excessive grazing by the essential oil content of its herbage. Second, the dominant perennial grasses in the forage base reproduced largely from seeds. When excessive numbers of livestock utilized the sagebrush/grasslands, grass disappeared and shrubs increased. If the grasses were not given a chance to redevelop their carbohydrate reserves to permit flowering and seed production, their chances of survival were slim. Under pristine conditions, no concentrations of large herbivores grazed the native grasses. The regeneration system of the native grasses was adapted to occasional stand renewal under conditions of exceptional precipitation.

The nineteenth-century method of determining stocking rates for the new rangelands of the West was to increase stock until losses became unacceptable. The overgrazing caused by this system gave native grasses no chance to establish new seedlings, and they declined annually regardless of the year's growth potential. The native herbaceous vegetation failed to adapt to the rapidly changing stand renewal system. The speed of evolutionary change did not equal the rate of environmental degradation. Native shrubs responded favorably to the destruction of herbaceous vegetation, but the exploitation levels of the two life forms were sufficiently diverse that the shrubs could never completely preempt the released potential. This environment was thus open to the introduction of plants able to outcompete the native perennial grasses. The annual destructive grazing of the herbaceous vegetation put a premium on annual growth forms, especially annuals with breeding systems that responded rapidly to changing environments to permit the evolution of adapted progenies.

The need to produce and harvest hay had profound sociological influences as well. Cattle ranching started as a type of agriculture that required minimum labor. Indians of the Intermountain area became a major source of this labor, both as year-round cowboys and as seasonal hay hands.

The turn of the century was a time of change on the sagebrush/grasslands. The national forest and federally funded reclamation projects would come into being in the next decade. Alien weeds such as cheatgrass had not yet reached the degraded sagebrush ranges. The old guard—Sparks, Harrell, Cleveland, Bradley, and Hardesty—was passing away. The cowboys who came astride horses and tried to avoid labor on foot were disappearing. Those who survived became ditch muckers and hay hands aside from their work with cattle and horses. The ranch hand—cowboy who came to town to drink and carouse was out of step with changing standards of social behavior.

The open-range era of livestock lasted only a brief time in its pure form. But the image of the cowboy as a free-living spirit of the plains became fixed in the imagination of the world and will forever symbolize the American West.

The last summer of the twentieth century proved to be one of those crossroads in history when time and space align to create an environmental disaster. In parts of northern Nevada the winter and spring of 1997 and 1998 provided excellent growing seasons for the exotic cheatgrass. Many reports told of cheatgrass two feet or even approaching three feet in height. Wildfire-suppression agencies held their collective breaths the entire summer, but the circumstances required for widespread nearly simultaneous ignition of wildfires did not occur. The winter and spring of 1998 and 1999 were not as favorable for the growth of cheatgrass, but an above-average crop grew beneath and among the weathered herbage from the previous season. Cheatgrass herbage carried over from the previous season dries much earlier in the season and provides a more compact fuel than that of the current season's production.

The stage was set for disaster in late July 1999 when weather conditions produced a line of thunderstorms across northern Nevada. The clouds were high and dry with an abundance of lightning. Well over one hundred nearly simultaneous lightning strikes ignited fires. The actual number will never be known because separately ignited fires burned to-gether into larger ones before firefighters arrived. The wildfire-suppression forces of public agencies were completely overwhelmed. When the smoke finally cleared and all the fires were out, estimates of the area burned ranged from 1 to 2 million acres, with 1.6 million acres the most commonly used figure. The cost of suppression was estimated at thirty-eight million dollars. This made the 1964 wildfires in Elko County, Nevada, where the term "firestorm" originated, look like bonfires by comparison.

The magnitude of the environmental disaster created a storm of public protest. The most common complaint was that federal firefighters, mostly U.S. Department of Interior, Bureau of Land Management (BLM) personnel, had not done their job. A backcountry cowboy in Elko County told J. A. Young later that summer that the BLM-ers were "real good at chasing those fires." Such appraisals were grossly unfair to the firefighters on the ground. The desert wildfire fighters have proven their courage many times. In the suburban Reno-Sparks–Carson City area of western Nevada they have repeatedly made a stand at an isolated home-site, saved the house, and then madly driven ahead of the flames to do the same thing over again. The men and women who spend a day and night taking risks like that can stand at the bar in any cowboy honky-tonk. Perhaps the magnitude of the firestorms in 1999 made it impossible for the agencies to suppress them. And better to chase the fires than to run before them and die. It was a tribute to the fire-suppression agencies that no one was killed during the 1999 firestorms in Nevada.[1] The rules of engagement under which the firefighters fought the fires limited their ability to suppress the fires but at the same time protected them from undue risk. Perhaps an analogy to a military unit engaged in warfare is appropriate. A squad can fight a battle with an enemy squad and over-come them through the application of proper training, appropriate high-quality equipment, and sheer bravery. A successful army needs a functioning command structure and a clear idea of how and why the battle must be won.

The environment and fire-suppression technology have both changed since the 1960s. During the 1964 firestorms in Elko County, airplanes were widely used in a largely uncoordinated attack, with planes flying crossing patterns in the smoke with limited communication facilities. The planes of 1999 were larger and more numerous and featured huge technological improvements in the type of fire retardant used and the method for dropping it. The pilots had not changed; they were still unbelievably skilled and courageous. Young and his wife sat on their patio in southwest Reno and waved to the pilots of four prop-driven planes as they repeatedly dropped into a canyon west of Hunter Creek to help suppress a lightning strike in dry grass above the Truckee River. High above, a twin-engine airplane circled to coordinate the aerial attack of the big planes while a few miles to the east commercial airliners continued their regularly scheduled landings and takeoffs at Reno-Tahoe Airport. One by one, the big tanker planes came down off the high ridge of the Carson Range of the Sierra Nevada in a long, slow glide. The silver wings and fuselage with fluorescent orange markings dropped lower and lower until the plane disappeared below the ridge that marked the small canyon only a half mile away. An observer's heart could not help but skip a beat as the plane disappeared from view. After an agonizing wait, the silver nose would reappear trailing a cloud of bright orange retardant and watchers could breathe again.

The biggest change in firefighting on wildlands since the 1960s has been the use of helicopters, which not only supply transportation and allow observation, but also play a huge role in actual fire suppression. Buckets suspended beneath them dip water from natural or artificial lakes and ponds or from portable plastic "swimming pools" refilled from ground tankers and carry it to fires.

How did the desert ranges of northern Nevada get into the environmental condition that prompted this terrible new chapter in wildfires? In earlier chapters we traced the origins of the livestock industry in the central Great Basin. Ranches were established by herdsmen without previous experience in temperate or cold deserts where periodic droughts and infrequent hard winters came with brutal intensity. Ranchers learned by experience.

By the end of the twentieth century, this was all in the past and "science" had supposedly come to dominate natural resource management. Grazing was managed, and the number of animals permitted to graze on the range was greatly reduced. The classes of livestock on the range had changed greatly, too. The range sheep industry that John Sparks fought so hard to exclude from his ranges had almost completely disappeared from northern Nevada. Free-roaming horses, protected from human predation by federal law, had greatly increased. Despite all that twentieth-century science could do, cheatgrass went from being a very minor component of the sagebrush/bunchgrass environment to the dominant element of millions of acres of rangeland. This dominance changed the chance of ignition and rate of spread, and prolonged the wildfire season.

During the 1960s federal land management agencies began a process that led to the establishment of grazing management systems on virtually all publicly owned rangeland in Nevada. From the end of World War II until the advent of widespread grazing management, it had been the policy of federal land management agencies to improve degraded rangelands by seeding perennial grasses. Federal management systems usually include some form of rest from grazing on a rotational basis. A common design is a three-pasture system in which one pasture is rested for an entire growing season from grazing. The year before it is rested the pasture is deferred from grazing until after seeds of the key forage species are ripe. The year before that, it is heavily grazed. This rest-rotation system is designed to favor the recruitment of seeds of perennial grasses to the rangelands being managed. Grazing after seeds ripen supposedly shatters the ripe seeds and helps to incorporate them into the seedbed.

Seedlings derived from this action are kept free of grazing for their first year.

The passage of federal environmental laws during the 1960s brought democracy to the management of publicly owned rangelands in the form of public review of actions that affect the environment. Protest from environmentalists that seeding perennial grasses on rangelands favored livestock at the expense of other range inhabitants was a major contributing factor to the end of range improvement practices and the shift to reliance on grazing management to restore perennial native plant dominance on Great Basin ranges.

There is no official scorecard for the success or failure of rest-rotation grazing systems on Nevada rangelands. Despite being the most extensive land management treatment ever applied to sagebrush/bunchgrass rangelands, it has never been comprehensively evaluated. Observations indicate that the results of nearly forty years of rest-rotation grazing management are highly site specific. At higher elevations where more native perennial grasses remained when the grazing management systems were initiated and where more environmental potential existed for supporting plant growth, the results have often been spectacularly successful, with return to dominance by native perennial grasses. This change has been so pronounced that wildfires in these perennial grasslands now occur in a manner similar to what must have occurred before European contact.

At lower elevations where fewer native perennial grasses remained when grazing management systems were established and the potential for plant growth was much lower, the results have been a dismal failure. Rest-rotation grazing systems under these conditions resulted in increased abundance and dominance of cheatgrass. If there were insufficient perennial grasses left to produce seed, there was no recruitment. And recruitment was virtually impossible in the face of competition from cheatgrass for the scant moisture available for plant growth. On the year grazing was deferred until after seed ripening and on the year of rest,

rest-rotation grazing management resulted in huge accumulations of cheatgrass herbage that constituted a severe wildfire hazard.

Ranchers pointed to the accumulations of cheatgrass at low elevations under grazing management systems and protested that grazing this herbage would both feed their animals and lessen the chance for wildfires. There is a considerable measure of truth in that, but grazing advocates should remember the three years of severe drought during the late 1980s when virtually no cheatgrass grew at low elevations in northern Nevada. On the vast areas of foothills range where cheatgrass had become the landscape-dominant species, there was virtually no living vegetation for three years. This was a rangeland disaster of much greater extent and magnitude than the wildfires of 1999 and 2000, but it received hardly any notice from range managers or the public.

Burgess L. Kay of the University of California collected data on cheatgrass production on a former big sagebrush/blue bunch wheatgrass site on the Likely Table in northeastern California.[2] The data he gathered indicate that there were more years when cheatgrass herbage production was below average than above. Precipitation "averages" for a given site are the result of many years with below-average precipitation, occasional very dry years, and very widely spaced years with much-above-average rainfall. It is the years with extremely favorable rainfall that produce record herbage production and the subsequent firestorms.

We have to know if the basic functions of a stable ecosystem can be sustained in a grazing system based on cheatgrass. Such processes as nutrient cycling and net carbon fixation may or may not be stable in a long-term grazing system based on cheatgrass. The extreme variability in the production of cheatgrass herbage is difficult to accommodate in a livestock production system based on cow and calf production. It does not mean that grazing to reduce the hazards of cheatgrass fuel production is impossible, but considerable thought and experimentation will be required to make such a system functional.

For the first three-quarters of the twentieth century cheatgrass became an increasingly apparent problem in former big sagebrush/bunch-grass potential plant communities. The lower margin of these sagebrush/bunchgrass potential plant communities adjoined the salt deserts. The ecotone between the two vegetation types was usually rather distinct, but on occasion big sagebrush would extend down into the salt deserts along seasonal watercourses or on sand-textured soils. Generally, the lower end of the big sagebrush communities very distinctly marked the end of the distribution of cheatgrass. The upper portions of the salt deserts often do not have salt-affected soils that limit the distribution of sagebrush, but the amount of precipitation they receive is so low that sagebrush cannot grow and is replaced by saltbush species and related shrubs. This relationship was so obvious that it became a truism that cheatgrass did not grow in the salt deserts of the Great Basin. Gradually in the 1980s, however, it became common to observe occasional cheatgrass plants in shadscale-dominated plant communities or around big sagebrush plants growing on sand in salt desert environments. Cheatgrass was not generally accepted as a species of the salt deserts until 1985, when huge wildfires occurred in salt desert environments near Winnemucca, Nevada, and on the California-Nevada border near the ghost town of Flanigan. The main wildfire northwest of Winnemucca, known as the Jungo Flat fire, reportedly advanced on a twenty-mile front at an estimated ten miles per hour. That works out to roughly 2,000 acres burning per minute. The truism concerning no cheatgrass in the salt deserts went up in smoke very rapidly. In the Winnemucca area about 500,000 acres burned in 1985.

In the BLM's Winnemucca District, about 800,000 acres of rangeland burned in the big firestorms of 1999. About one-half of this acreage occurred in salt desert–type plant communities. If you cross the Rye Patch dam on the Humboldt River between Lovelock and Winnemucca and drive northwest into the area known as Poker Brown Gap, there is a vast expanse of some 240,000 acres of burned shadscale salt desert vegetation converted to cheatgrass dominance by wildfire. The western salt deserts have always been considered free of wildfires, and catastrophic fires were thus not viewed as a form of stand renewal in such habitats. It is a principle of plant ecology that the way an existing plant community is renewed—or in the case of wildfires, destroyed—determines the characteristics of the next community that occupies the site. The introduction of cheatgrass to the salt deserts of the Great Basin brought fire as a stand-renewal process and set off a complicated series of actions and interactions that no one yet fully understands.

Virtually no one asks why cheatgrass suddenly spread into the upper portion of the salt desert environments. Was a new type of supercompetitive cheatgrass accidentally introduced into North America? Did the cheatgrass already established in the higher-environmental-potential sagebrush zone hybridize and through natural selection evolve a new form adapted to the aridity of the salt deserts? Perhaps the invasion was the result of climatic change too subtle to be readily apparent in weather records. The reduction of the range sheep industry that formerly wintered on the salt desert ranges may have contributed to this multi-million-acre invasion.

The native perennial grasses vary in their responses to wildfires. Before cheatgrass was introduced, such stands burned late in the summer or early fall, when the grasses had matured and dried sufficiently to carry a fire. Under those conditions, probably none of the perennial grasses were injured by the wildfires. The fires served to eliminate woody species and favored the grasses. Cheatgrass matures in early summer, advancing the fire season into the period when the native perennial grasses are still growing. Some species are more susceptible to wildfire damage than others under these conditions. The most destructive thing that can happen to the perennial grasses burned in a wildfire is to have their herbage heavily grazed the first season after burning. Domestic and wild herbivores are attracted to the lush herbaceous vegetation that sprouts in burns because of reduced competition and more available nitrogen. The nitrogen may result in a higher protein content of the herbage, which also attracts the herbivores. If the burned area is grazed, the native per-

ennial grasses never have a chance to take advantage of the reduced competition and available nitrogen. If the grazing occurs after the thin stand
of postburn cheatgrass is mature, animals will selectively graze every
green perennial grass. Federal land management agencies have therefore decreed that domestic livestock may not graze areas burned in
wildfires for two years after the fire. This means the rancher has to find
alternative forage for two years or sell at least a portion of his brood stock.

The impact of such conservation-designed policy is directly related
to the extent and distribution of the area burned. The 500,000 acres of
rangeland that went out of use after the fires near Winnemucca, Nevada,
in 1985 had a significant negative influence on the local agricultural
economy. Generally, the area affected is larger than the actual area
burned. If 2,000 acres in a 6,000-acre pasture burn, the entire 6,000
acres is off-limits for grazing for the two-year period. It is seldom economically feasible to temporarily fence off the burned acreage from the
remaining unburned area. This is especially true if essential resources
for livestock, such as watering points, are located predominantly in the
burned area.

The policy of deferring grazing for two years after rangelands are
burned in wildfires is generally very unpopular with ranchers. The more
knowledgeable ranchers point out that some fragile range sites should
be protected from grazing for perhaps five years after burning, but sites
that are completely dominated by cheatgrass with few or no native perennial grasses become herbaceous wildfire fuel bombs during a two-year
exclusion from grazing. This is an excellent example of the relentless
application of an ecological principle in direct opposition to common
sense.

During the first half of the twentieth century, cheatgrass on the Nevada rangelands was biologically suppressed by excessive grazing and by
the increase in sagebrush that occurred after the native perennial grasses
were reduced. After World War II, a series of interacting events favored
cheatgrass over the native perennials. The range sheep industry greatly

declined for a variety of reasons. Federal land management agencies made a concerted effort to improve the quality of range management. Community allotments were subdivided to single-permit allotments, making one permittee responsible for the range condition of his or her allotment. The number of livestock allowed on each allotment and its season of use were adjusted to bring the numbers of livestock grazed into closer agreement with the carrying capacity of the ranges. All of these measures were well intentioned and theoretically conceived to help return the native perennial species to dominance. Unfortunately, the ecological threshold of Nevada rangelands is determined by a combination of the land's environmental potential and its successional stage. If a given site is above this threshold and the grazing pressure and season of use are properly regulated through management, succession proceeds to perennial grass dominance. Below the threshold, reduced grazing pressure results in a cyclic, explosive expression of cheatgrass. Once cheatgrass dominance has been expressed, the way is open for catastrophic wildfires that remove the sagebrush and accelerate cheatgrass dominance.

The widespread seeding of exotic perennial grasses on Nevada rangelands after World War II greatly increased the harvestable forage base and reduced the grazing pressure on native perennial bunchgrass ranges. Again the ecological threshold came into play. Higher-potential sites that had a remnant stand of perennial grasses responded favorably. Lower-potential sites without a critical mass of perennials regressed to cheatgrass dominance. The complexity of ecological relations in sagebrush/bunchgrass communities is well illustrated in the exotic perennial grass seeding program. The original seeds faced competition from sagebrush. The rangeland plow was developed to destroy old-growth stands of big sagebrush that lacked a perennial grass understory. The technologies that were developed were very successful.[3] About one million of the nineteen million acres of sagebrush/bunchgrass potential rangelands were converted to transitory exotic perennial grass dominance through these programs. During the last days of this golden age of range improvement, competition from cheatgrass made artificial seeding of degraded sagebrush rangelands increasingly difficult. Cheatgrass was increasing because of reduced grazing pressure, which was substantially due to the success of the exotic perennial grass-seeding program. What a paradox: successful seeding of perennial grasses helped create an environmental situation in which weed competition limited the further establishment of perennial grass seeds by reducing overgrazing.

The federal land management agencies' answer to the firestorms of 1999 has been to seed the burned ranges to suppress cheatgrass. On one day in 1999, the Bureau of Land Management purchased nearly five million pounds of seed to plant on Nevada ranges. If you have the feeling that you have just been led around in a circle, it is a perfectly reasonable assumption. The cost for environmental restoration was greater than the cost of fire suppression in 1999. None of the restoration seeding is designed to produce livestock forage; it is officially conducted to prevent future wildfires, prevent accelerated erosion, and restore wildlife habitat.

The suppression of wildfires and the restoration of burned rangelands in the Great Basin is big business. An Idaho rancher who belongs to the family that now operates John Sparks's famous Winecup field on Goose Creek described the federal fire-suppression efforts on the family's grazing land as "the circus came to town." The local store, restaurant, and motel made a killing. The town paid off the used fire truck it had bought on credit. All it had to do was park it by potentially endangered structures. Eleven track-laying tractors equipped with bulldozer blades and their operators were delivered to the fire site, but they never moved because archaeological technicians were not available to scout in front of the blades for cultural artifacts. It was the middle of a terrible wildfire season across the West, and the highly trained hot shot crews were already deployed on more pressing fires. The crews that were available lacked training and experience to the point that the fire bosses were afraid to deploy them for safety considerations. Many years ago, there was an old BLM employee in the agency state headquarters in Reno. Bill was a bachelor, and every year he volunteered to serve on overhead crews on big wildfires. He became widely recognized as an expert fire boss with experience from Alaska to Florida and across the western range. His guiding principle in fire suppression was that all past wildfires have gone out. The Bill Principle eventually applied to the Goose Creek fire; it went out. The circus left town, and the rancher was left without forage for his livestock for two years.

Congress decided in 2000 that the answer to the wildland fire problem in the West was a huge increase in funding for fire-suppression equipment and more firefighters. The disease is degraded range plant communities dominated by an exotic annual grass. The symptom is the resulting wildfires. The symptoms will disappear if the disease is cured through environmental restoration that includes grazing domestic livestock as a proper management tool.

The environment of the cold desert rangelands is not a simple place where complex phenomena are easily explained. It is a place of trails that have grown dim since the hooves of the Longhorns John Sparks drove from Texas first found their way among the gray sagebrush and bunchgrass. Modern society is environmentally aware but not necessarily willing to expend the time and energy required to unravel the dim trails of environmental history. It is much simpler to take a shortcut and place the blame for environmental ills and potential restoration on single factors. "Graze the cheatgrass, and do not let it burn." "Reave all domestic livestock from the public range, and the cheatgrass will disappear." These are the conflicting rallying cries of highly vocal factions of the polarized society that wants to impose policy on the management of the western range.

John Sparks and his associates brought cattle to the cold deserts because the desert ranges were lands no one else wanted late in the nineteenth century. He built an empire based on cattle. The Nevada portion of his holdings equaled 3 percent of the state's land area. Despite the extent of his ranches, John Sparks and his fellow livestock pioneers actually owned very little of the desert landscape. The federal government never evolved a policy that permitted passage from the public domain of the vast acreage necessary for a cold-desert ranching empire. Obviously, it was possible to control such an empire through owning hay land and water rights. Starting on the eastern seaboard and moving westward, the wilderness was developed and passed into private ownership. Only in the arid West and the Arctic vastness of Alaska does the federal government remain the landlord of significant portions of the landscape. Uncle Sam owns 87 percent of Nevada. In the counties of central Nevada the federal government owns virtually all the land. In the fall of 2000 a convention was held in Reno dedicated to the principle that Americans had no right to graze livestock, cut mature trees for conversion to lumber, develop mines for mineral extraction, or drill for oil and gas on lands the federal government retains through the lack of a valid land-disposal policy. These lands are to be locked in the control of a federal bureaucracy determined to restore them to the conditions that existed at the time of European contact. For much of the history of the United States romantic idealists have believed in the Jeffersonian concept that only a society based on small farmers can support democracy.

In a highly ironic twist, the Jeffersonian vision of small landholdings in a rural atmosphere is actually unfolding on the landscape of northern Nevada. Nevada was among the leading states in population growth during the last decades of the twentieth century, but this growth occurred largely in the Las Vegas and Reno metropolitan areas. The population increase in Nevada and California has made some people long to escape from suburbia. The sale of the railroad grants across northern Nevada (9 percent of the state) provides an outlet for those who desire escape to the wilderness. The desert homesteads often do not have access to electric power grids, a well, or sewage disposal; the factory-built house sits on concrete blocks at the end of a rocky dirt track in the wilderness. One

desert traveler paraphrased the old derogatory cowboy name for homesteaders, "nesters," to "messters" to describe this new wave of settlers
and their propensity to accumulate well-used but potentially useful symbols of civilization around their tin nests. The spread of these desert
homesites along the railroad checkerboard across northern Nevada has
created a potential nightmare for wildland firefighters when the next big
firestorm occurs. The prime directive—protect human life on isolated
homesites—may result in even bigger acreages being burned.

Despite the gloom and doom that abounds with the cheatgrass-fueled
wildfires in the foothills, the rangelands of northern Nevada were probably in better ecological condition in 2001 than they were in 1901. The
upland range areas have greatly benefited from the introduction of grazing management systems. The critical area is the foothills, where Wyoming big sagebrush has actually come close to elimination over vast areas. The central question is, can cheatgrass-dominated rangelands
sustain livestock? From a grazing standpoint, the cheatgrass-invaded
salt desert areas have more forage than existed before the invasion.

Most Americans have never known anything but a superabundance of
high-quality, cheaply priced food. Worldwide, however, high-quality
protein is the nutrient most often lacking in human diets. The futurists
tell us that the increasing human population will create a severe shortage of food before we are very far into the twenty-first century. Americans may very well have to forgo our feedlot-fattened beef, poultry, and
pork because all of these livestock production systems require feed that
can be directly consumed and digested by humans. Range livestock production relies on the use of ruminant animals to convert herbage so high
in cellulose that humans cannot digest it directly. The range livestock
industry produces quality protein from plant material that will not support human life. Ranges are the direct link to much of western North
American history. They also may be the future for America if they are
managed on a sustainable basis.

PROLOGUE

1. D. Morgan, *The Humboldt Highroad of the West* (New York: Farrar and Rinehart, 1943).

2. R. Butterfield, "Elko County," *Life* 76, no. 16 (April 18, 1949): 98–114.

3. Unless otherwise cited, information is from testimony recorded at the fire critique held by the Bureau of Land Management in Elko and on file in the BLM state office in Reno. See clippings from *Chicago Tribune, Los Angeles Times, San Francisco Chronicle, Deseret News* (Salt Lake City), and *Burlington Free Press* (Burlington, Vermont) for August 16–20, 1964.

4. Personal communication to James A. Young from Dr. Raymond A. Evans, Investigation Leader, Pasture and Range Management Project, USDA, Agricultural Research Service, Reno, Nevada.

5. Personal experience of James A. Young, Siskiyou County, California, 1955.

6. *Reno Evening Gazette*, August 20, 1964; and BLM records.

7. I. C. Russell, *Geological History of Lake Lahontan* (Washington, D.C.: Government Printing Office, 1885). [Government Printing Office hereinafter referred to as GPO.]

8. P. B. Kennedy, *Summer Ranges of Eastern Nevada Sheep*, Nev. Agric. Exp. Sta. Bull. 55 (Reno: University of Nevada, 1903).

9. Hot shot crews are organized by various fire-suppression agencies, given special training and equipment, and used as shock troops to knock down fires as soon as they are reported and before they get out of control. To add to their esprit de corps, these crews often give themselves names and stitch them on their Levis jackets.

10. L. W. Mills, *A Sagebrush Saga* (Springville, Utah: Art City Publishing, 1954).

11. J. A. Young and R. A. Evans, "Population Dynamics after Wildfires in Sagebrush Grasslands," *J. Range Manage.* 31 (1978): 283–89. Much of the 300,000 acres that burned in the 1964 fire was restored by a superb revegetation effort conceived by J. Russell Penny and directed by Bill Malencik. For details, see Guy R. Sheeter, "Secondary Succession and Range Improvement after Wildfires in Northeastern Nevada" (master's thesis, University of Nevada, Reno, 1968).

1. GRAY OCEAN OF SAGEBRUSH

1. "Comments on Travel on the Oregon Trail." See John Townsend, *Narrative of a Journey across the Rocky Mountains to the Columbia River,* in *Early Western Travels,* ed. R. Thwaites, 21:107–639 (Cleveland: Arthur H. Clark, 1906). See also F. Wisenzenius, *A Journey to the Rocky Mountains in the Year of 1839* (St. Louis: Missouri Historical Society, 1912); and J. C. Frémont, *Report of the Exploring Expedition to the Rocky Mountains in the Year 1842, and to Oregon and Northern California in the Years 1843–44* (Washington, D.C.: Gales and Seaton, 1845).

2. For readability, scientific names have been kept to a minimum. An equivalency table of common and scientific names for plants and animals precedes the index.

3. George Stewart, "The History of Range Use," in *The Western Range,* S. Doc. 199, 74th Cong., 2nd sess. (Washington, D.C.: GPO, 1936); and T. R. Vale, "Presettlement Vegetation in Sagebrush/Grassland Areas of the Intermountain West," *J. Range Manage.* 28 (1975): 32–36. For further information, see George Stewart, "Historic Records Bearing on Agricultural and Grazing Ecology in Utah," *J. Forestry* 39 (1941): 363–75. For the reconstruction of the vegetation of a specific valley, see A. C. Hull and M. K. Hull, "Presettlement Vegetation of Cache Valley, Utah and Idaho," *J. Range Manage.* 27 (1974): 27–29.

4. See references in note 1, above.

5. E. W. Tisdale, J. Hironaka, and M. A. Foshers, *The Sagebrush Region in Idaho,* Agric. Exp. Sta. Bull. 512 (Moscow: University of Idaho, 1969).

6. Alexander Ross, *The Fur Hunters of the Far West,* ed. M. M. Quaite (Chicago: Lakeside Press, 1924).

7. P. S. Ogden, *Peter Skene Ogden's Snake Country Journals, 1824–1825 and 1825–1826,* ed. E. E. Rich and H. M. Johnson (London: XIII Hudson's Bay Record Society, 1950). For an overview of exploring the Great Basin, see G. G. Cline, *Exploring the Great Basin* (Norman: University of Oklahoma Press, 1963).

8. Personal communications to James A. Young from Mary Rusco, Nevada State Museum, Carson City.

9. W. T. Hornaday, *The Extermination of the American Bison,* U.S. Nat. Hist. Mus. Rep. 1887 (Washington, D.C.: GPO, 1889). For additional information, see C. H. Merriam, "The Buffalo in Northeastern California," *J. Mammal.* 7 (1926): 211–14.

10. B. R. Butler, "The Evolution of the Modern Sagebrush-Grass Steppe Brome on the Eastern Snake River Plain," in *Holocene Environmental Change in the Great Basin*, Nev. Archaeol. Soc. Res. Pap. 6, ed. Robert Elston, 5–39 (Reno: Nevada Archaeological Society, 1976).

11. M. H. Egan and H. R. Egan, *Pioneering the West, 1846 to 1878*, ed. W. M. Egan (Salt Lake City: privately printed, 1917).

12. E. R. Hall, *Mammals of Nevada* (Berkeley: University of California Press, 1946). See also J. K. McAdoo and J. A. Young, "Jackrabbits," *Rangelands* 2 (1980): 135–38.

13. A. W. Sampson, "Succession as a Factor in Range Management," *J. Forestry* 15 (1917): 593–96.

14. H. M. Hall and F. E. Clements, *The Phylogenetic Method in Taxonomy: The North American Species of* Artemisia, Chrysothamnus *and* Atriplex (Washington, D.C.: Carnegie Institute, 1923). Original citation in Thomas Nuttall, *Trans. Amer. Phil. Soc.* 2 (1841): 398.

15. A. A. Beetle, *A Study of Sagebrush*, Wyo. Agric. Exp. Sta. Bull. 368 (Laramie, 1960).

16. S. F. Arno and K. M. Sneck, "A Method for Determining Fire History in Coniferous Forest and Range" (Ogden, Utah: USDA, Forest Service Experiment Station, 1977).

17. J. A. Young and R. A. Evans, "Wildfires in Semiarid Grasslands," in *Proceedings of the XIII International Grassland Congress*, Leipzig, GDR, 1977. See also J. A. Young and R. A. Evans, "Population Dynamics of Green Rabbitbrush in Disturbed Big Sagebrush Communities," *J. Range Manage.* 27 (1974): 127–32.

18. H. A. Wright, L. L. Neuenschwander, and C. M. Britton, "The Role and Use of Fire in Sagebrush-Grass and Pinyon Juniper Plant Communities" (Ogden, Utah: USDA, Forest Service Experiment Station, 1978).

19. A. H. Winward and E. W. Tisdale, *Taxonomy of the* Artemisia tridentata *Complex in Idaho*, Coll. Forest. Bull. 19 (Moscow: University of Idaho, 1977).

20. J. G. Nagy, H. W. Steinhoff, and G. M. Ward, "Effect of Essential Oils of Sagebrush on Deer Rumen Microbial Functions," *J. Wildl. Manage.* 28 (1964): 784–90.

21. J. G. Nagy, "Wildlife Nutrition and the Sagebrush Ecosystem," in *Proceedings of the Sagebrush Ecosystem: A Symposium*, 144–68 (Logan: Utah State University, 1978).

22. Jeff Powell, "Site Factor Relationships with Volatile Oils in Big Sage-brush," *J. Range Manage.* 23 (1970): 42–46. See also D. L. Hanks, J. R. Brunner, D. R. Christensen, and A. P. Plummer, "Paper Chromatography for Determining Palatability Differences in Various Strains of Big Sagebrush," Res. Pap. INT-101 (Ogden, Utah: USDA, Forest Service Experiment Station, 1971).

23. E. H. Cronin, Phil Ogden, J. A. Young, and William Laycock, "The Ecological Niches of Poisonous Plants in Range Communities," *J. Range Manage.* 31 (1978): 328–34.

24. C. W. Ferguson and R. R. Humphrey, "Growth Rings of Sagebrush Reveal Rainfall Records," *Progressive Agric.* (summer 1959): 3.

25. Beetle, *A Study of Sagebrush.* See also Humboldt Basin statistics from U.S. Soil Conservation Service, *Water and Related Land Resources, Humboldt River Basin, Nevada,* Rep. 12 (Carson City.: USDA, 1966). Information on sagebrush in northeastern California is from J. A. Young, R. E. Evans, and Jack Major, "Sagebrush Steppe," in *Terrestrial Vegetation of California,* ed. M. G. Barbour and Jack Major, 763–96 (New York: John Wiley and Sons, 1977).

26. Ben Zamora and P. T. Tueller, "*Artemisia arbuscula, A. longiloba,* and *A. nova* Habitat Types in Northern Nevada," *Great Basin Nat.* 33 (1973): 225–42.

27. Young et al., "Sagebrush Steppe." Also see J. F. Franklin and C. T. Dryness, *Natural Vegetation of Oregon and Washington,* Pac. Northwest Forest and Range Exp. Sta. Tech. Rep. NW-8 (Portland, Ore.: USDA 1973).

2. THE EXPLOITATION PAGEANT

1. Byron D. Lusk, "Golden Cattle Kingdoms of Idaho" (master's thesis, Utah State University, Logan, 1978).

2. Washington Irving, *Astoria* (New York: G. P. Putnam, 1849).

3. I. C. Russell, *Geology and Water Resources of the Snake River Plains of Idaho* (Washington, D.C.: GPO, 1902).

4. In order to encompass a single day, this description is a composite of I. C. Russell's description.

5. D. I. Axelrod, "The Miocene Trapper Creek Flora of Southern Idaho," *Geol. Sci.* 51 (1964): 371–92.

6. Lusk, "Golden Cattle Kingdoms."

7. J. C. Frémont, *Report of the Exploring Expedition to the Rocky Mountains* (Ann Arbor: University Microfilms, 1977).

8. H. R. Cramer, *Hudspeth's Cutoff* (Burley, Ida.: privately printed, 1969).

9. Lusk, "Golden Cattle Kingdoms."

10. *Idaho Sunday Statesman* (Boise), February 2, 1941.

11. A. C. Anderson, ed., *Trails of Early Idaho: The Pioneer Life of George Goodhart* (Caldwell, Ida.: Caxton, 1940).

12. D. O. Hyde, *The Last Freeman* (New York: Dial Press, 1973).

13. M. F. Schmitt, ed., *The Cattle Drives of David Shirk* (Portland, Ore.: Champoeg Press, 1956).

14. Adelaide Hawes, *The Valley of Tall Grass* (Bruneau, Ida.: privately printed, 1950).

15. "Mimeograph History of Minidoka National Forest" (Twin Falls, Ida.: USDA, Forest Service, n.d.). Also see Alexander Toponce, *Reminiscences of Alexander Toponce* (Norman: University of Oklahoma Press, 1971).

16. C. W. Gordon, "Report on Cattle, Sheep and Swine," *Supplementary to Enumeration of Livestock on Farms in 1880, Tenth Census of the United States*, vol. 3 (Washington, D.C.: GPO, 1880).

17. Schmitt, *Cattle Drives of David Shirk*.

18. Gordon, "Report on Cattle, Sheep and Swine."

19. Morgan, *The Humboldt Highroad of the West*.

20. Information on Bill Downing is from the narrative of Captain C. H. Roland, Wells, Nevada, on file with Nevada Historical Society. The H-D Ranch was located at the site of the present-day Winecup Ranch.

21. E. B. Patterson, L. A. Ulph, and Victor Goodwin, *Nevada's Northeast Frontier* (Sparks, Nev.: Western Printing and Publishing, 1969). See also V. S. Truett, *On the Hoof and Horn in Nevada* (Los Angeles: Gehrett-Truett-Hall, 1950).

22. J. M. Townley, "Reclamation in Nevada, 1850–1904" (Ph.D. diss., University of Nevada, Reno, 1976).

23. *Elko Independent*, December 15, 1864.

24. J. M. Guinn, *History of the State of California and Biographical Record of the San Joaquin Valley* (Chicago: Chapman, 1905).

25. Personal communication to James A. Young from N. J. Harrell, Louis Harrell's son, Twin Falls, Ida., July 1979.

26. C. S. Walgomott, *Reminiscences of Early Days*, 2 vols. (Twin Falls, Ida.: privately printed, 1927).

27. N. L. Bowman, *Only the Mountains Remain* (Caldwell, Ida.: Caxton, privately printed, 1955).

28. Townley, "Reclamation in Nevada."

29. J. O. Oliphant, *On the Cattle Ranges of the Oregon Country* (Seattle: University of Washington Press, 1968).

30. C. M. Owens, "Early Cattle Raising in Wyoming" (master's thesis, Colorado State Teachers College, Greeley, 1932).

3. LEFT-HAND TRAIL TO HELL

1. Russell, *Geological History of Lake Lahontan.*

2. Ibid.

3. S. G. Houghton, *A Trace of Desert Waters* (Glendale, Calif.: Arthur H. Clark, 1976).

4. W. A. Dayton, *Important Western Browse Plants*, USDA Misc. Bull. 101 (Washington, D.C.: GPO, 1931).

5. *Range Plant Handbook* (Washington, D.C.: GPO, 1937).

6. H. C. Stutz, "Explosive Evolution of Perennial *Atriplex* in Western America," in *Intermountain Biogeography: A Symposium,* Great Basin Nat. Mem. 2, 161–68 (Provo, Utah: Great Basin Naturalist, 1980); H. C. Stutz, J. M. Melhy, and G. K. Livingston, "Evolutionary Studies of *Atriplex*: A Relic Gigas Diploid Population of *Atriplex canescens,*" *Amer. J. Bot.* 62 (1975): 236–48.

7. Personal communication to James A. Young from Frosty Tipon of the T Quarter Circle Ranch in Winnemucca, Nevada. This ranch winters cattle on the sand dunes of Silver State Valley.

8. J. G. Smith, *Fodder and Forage Plants Exclusive of the Grasses,* USDA, Div. Agrostol. Bull. 2 (Washington, D.C.: GPO, 1900).

9. W. D. Billings, "The Shadscale Vegetation Zone of Nevada and Eastern California in Relation to Climate and Soil," *Amer. Midl. Nat.* 42 (1949): 87–109.

10. A. Nelson, *The Red Desert of Wyoming and Its Forage Resources,* USDA, Div. Agrostol. Bull. 13 (Washington, D.C.: GPO, 1892).

11. L. A. Stoddart and A. D. Smith, *Range Management* (New York: McGraw-Hill, 1943). Cows require 7.5–12 gallons of water per day, and sheep need 0.2–1.5. Generally, five sheep are considered equal to one cow in forage requirements, making the sheep's maximum water requirement the same as the minimum cow requirement.

12. *Range Plant Handbook.*

13. I. C. Russell, *Quaternary History of Mono Valley, California,* USGS, Eighth Annual Report, 1886–87, pt. 1, pp. 261–394 (Washington, D.C.: GPO, 1889).

14. J. A. Young, R. A. Evans, and P. T. Tueller, "Plant Communities Pristine and Grazed," in *Holocene Environmental Change in the Great Basin,* Nev. Archaeol. Surv. Res. Pap. 6, ed. Robert Elston, 187–215 (Reno: University of Nevada, 1976).

15. R. E. Eckert Jr., *Improvement of Mountain Meadows in Nevada,* BLM Res. Rep. 4400 (Reno: U.S. Department of the Interior, 1975).

16. D. A. Klebenow, "The Habitat Requirements of Sagegrouse and the Role of Fire in Management," in *Proceedings of the Annual Tall Timbers Fire Ecology Conference* (Tallahassee: Tall Timbers Research Station, 1972); Lieutenant H. L. Abbott, *Explorations for a Railroad Route from the Sacramento Valley to the Columbia River,* Ex. Doc. 91, 33rd Cong., 2nd sess., House of Representatives (Washington, D.C.: A. O. P. Nicholson, 1855); D. E. Savage, "The Relationship of Sagegrouse to Upland Meadows in Nevada" (master's thesis, University of Nevada, Reno, 1968).

17. C. D. Marsh, A. B. Clauson, and H. Marsh, *Larkspur or Poison Weed,* USDA Farmer's Bull. 988 (Washington, D.C.: GPO, 1934).

18. Information on desert bighorns from Hall, *Mammals of Nevada.* The last bighorn sighted in Elko County was in the Ruby Mountains in 1921.

19. Hall, *Mammals of Nevada.* Quote is from J. H. Simpson, *Report of Exploration across the Great Basin of the Territory of Utah for a Direct Wagonroute from Camp Floyd to Genoa, in Carson Valley in 1859* (Washington, D.C.: GPO, U.S. Engineer Department, 1876).

20. Hawes, *The Valley of Tall Grass.*

4. TEXAS CATTLE AND CATTLEMEN

1. Gordon, "Report on Cattle, Sheep and Swine." For details of the Spanish influence, see also "The Spanish Approach to the Great Plains," in W. P. Webb, *The Great Plains* (Boston: Ginn and Company, 1931).

2. J. Frank Dobie, *The Longhorns* (Boston: Little, Brown, 1941).

3. T. G. Jordan, *Trails to Texas* (Lincoln: University of Nebraska Press, 1981).

4. J. G. McCoy, *Historic Sketches of the Cattle Trade of the West and Southwest,* ed. Ralph Bicher (Glendale, Calif.: Arthur H. Clark, 1940).

5. J. E. Rouse, *The Criollo-Spanish Cattle in the Americas* (Norman: University of Oklahoma Press, 1977).

6. Dobie, *Longhorns.*

7. Gordon, "Report on Cattle, Sheep and Swine."

8. Louis Pelzer, *The Cattleman's Frontier* (Glendale, Calif.: Arthur H. Clark, 1936).

9. L. F. Allen, *American Cattle—Their Breeding and Management* (New York: Orange Judd Company, 1890).

10. James MacDonald, *Food from the Far West, or American Agriculture, with Special Reference to Beef Production and Importation of Dead Meat from America to Great Britain* (London: William P. Nimmo, 1878).

11. Ibid.

12. Webb, *The Great Plains.*

13. Gordon, "Report on Cattle, Sheep and Swine."

14. Webb, *The Great Plains.*

15. McCoy, *Historic Sketches of the Cattle Trade.*

16. Webb, *The Great Plains.*

17. Ibid.

18. Dobie, *Longhorns.*

19. Snyder manuscripts on file in collection at Barker Historical Center, University of Texas Library, Austin. Unindexed material researched by Mrs. Joe B. Gordon.

20. Ibid.

21. Ibid. Colonel Dudley H. Snyder died on September 19, 1921, at the age of eighty-eight. He was a supporter of Southwestern University in Georgetown, Texas.

22. John Clay, *My Life on the Range* (Chicago: privately printed, 1923).

23. C. S. Scarbrough, *Land of Good Water, Takachue Pouetsu, a Williamson County, Texas, History* (Georgetown, Tex.: Williamson County Son Publishers, 1974).

24. Obituary file for John Sparks, Nevada Historical Society, Reno.

25. R. I. Fulton, "Camp Life on Great Cattle Ranches in Northern Nevada," *Sunset* 5, no. 3 (1900): 111–18.

26. *Seventh Census of the United States,* Records for White Township, Ashley

County, Arkansas (Washington, D.C.: GPO, 1850). Sparks never mentioned the family's move to Arkansas even though at least one of his sisters was born in White Township. Mrs. Joe B. Gordon traced the Sparks family from Mississippi to Arkansas.

27. *Eighth Census of the United States,* Records for Lampasas County, Texas (Washington, D.C.: GPO, 1860).

28. J. B. Barry, *The Days of Buck Barry in Texas, 1845–1906,* ed. J. K. Green (Austin: University of Texas Library Archives, 1936). Also see R. P. Felgar, "Texas in the War for Southern Independence, 1861–1865" (Ph.D. diss., University of Texas, Austin, 1935).

29. Information on Sparks's enlistment from Civil War records, University of Texas, Austin. Research by Mrs. Joe B. Gordon. See also *Prose and Poetry of the Livestock Industry of the United States,* prepared by the National Livestock Association (Denver: National Livestock Historical Association, 1904). The military service of John Sparks has been subjected to a great deal of distortion. See, for example, D. H. Grover, *Diamondfield Jack* (Reno: University of Nevada Press, 1968). Grover even assigned Sparks to the Union army.

30. John Sparks's personal history, dictated to Hubert Bancroft, Bancroft Oral History Program, Bancroft Library, University of California, Berkeley.

31. Ibid.

32. J. H. Triggs, *History of Cheyenne and Northern Wyoming* (Omaha: Herald Steam Book and Job Printing House, 1876).

33. A. W. Spring, *Seventy Years—A Panoramic History of the Wyoming Stock Growers Association* (Laramie: Wyoming Stock Growers Association, 1942).

34. Maurice Frink, *Cow Country Cavalcade* (Denver: Old West Publishing, 1954).

35. E. S. Osgood, *The Day of the Cattleman* (Chicago: University of Chicago Press, 1929).

36. Ibid., quoting statements of Representative James A. Garfield in *Congressional Record,* 43d Cong., 1st sess., pp. 2107–9.

37. *Climate and Man,* USDA Yearbook of Agriculture (Washington, D.C.: GPO, 1941).

38. C. M. Owens, "Early Cattle Raising in Wyoming" (master's thesis, Colorado State Teachers College, Greeley, 1932).

39. Sparks's personal history dictated to Bancroft. See also Wilkerson manuscript of unpublished biographical material at Wyoming State Archives and Historical Department, Laramie.

40. Owens, "Early Cattle Raising in Wyoming."

41. Maurice Frink, W. T. Jackson, and A. W. Spring, *When Grass Was King* (Boulder: University of Colorado Press, 1956).

42. Sparks's personal history dictated to Bancroft.

43. M. S. Yost, *The Call of the Range: The Story of the Nebraska Stock Growers Association* (Denver: Sage Books, 1966).

44. Scarbrough, *Land of Good Water.* An account sheet showing the Snyder brothers' balance in the Steele and Sparks Bank is preserved in the Snyder Papers, Barker Historical Center, University of Texas Library, Austin.

45. Ibid.

5. BUY, BEG, BORROW, OR STEAL A RANCH

1. B. H. Hibbard, *A History of Public Land Policies* (New York: Macmillan, 1924).

2. *United States Land Report, 1875–1876,* no. 1680 (Washington, D.C.: GPO, 1876).

3. Webb, *The Great Plains.*

4. Gordon, "Report on Cattle, Sheep and Swine." See also Thomas Donaldson, *The Public Domain: Its History with Statistics* (Washington, D.C.: GPO, 1889; reprint, Johnson Reprint Corp., 1970). Also see *Report from the Chief of the Bureau of Statistics, 1884–1885,* Ex. Doc. 267, 48th Cong., 2nd sess., House of Representatives (Washington, D.C.: GPO, 1885); hereafter cited as "Nimmo Report." The survey was in response to a House resolution calling for information on the range and ranch cattle industry in the western states and territories.

5. Major J. W. Powell, *Report on the Lands of the Arid Regions of the United States* (Washington, D.C.: GPO, 1878).

6. E. O. Wooten, *The Public Domain of Nevada and Factors Affecting Its Use,* USDA Tech. Bull. 301 (Washington, D.C.: GPO, 1932).

7. Based on data prepared by Hugh A. Shamberger, former Nevada state engineer, and held in the Shamberger Collection, Nevada State Historical Society. Summary of data presented in Townley, "Reclamation in Nevada."

8. Townley, "Reclamation in Nevada."

9. George Kraus, *High Road to Promontory* (Palo Alto, Calif.: American West, 1969).

10. Townley, "Reclamation in Nevada."

11. U.S. Bureau of Land Management, Record Group 49, Office of the U.S. Surveyor-General for Nevada, Federal Record Center, San Bruno, Calif.

12. "Biennial Report of the Surveyor-General and State Land Register of the State of Nevada for 1867 and 1868," in the *Journal of the Senate,* fourth session of the legislature of the State of Nevada, 1869, begun on Monday, January 4, and ended on Thursday, March 4 (Carson City: State Printer, 1869).

13. Townley, "Reclamation in Nevada." The fixation with artesian water was not limited to Nevada. The U.S. Geological Survey conducted groundwater surveys in many western states and territories during the late nineteenth century to determine the potential for artesian wells.

14. Townley, "Reclamation in Nevada."

15. Ibid. Townley based most of this discussion on letters found in the U.S. Bureau of Land Management, Record Group 49, Office of the U.S. Surveyor-General for Nevada, Federal Record Center, San Bruno, Calif. Specifically, Townley quotes a letter from E. D. Kelley to Thomas Short, March 21, 1900; and a letter from E. D. Kelley to H. H. Porch, March 10, 1900.

16. Ibid., specifically, letter from J. E. Jones to John Sparks, January 14, 1890; letter from H. D. Notware to Miller and Loy, September 14, 1891; letter from A. C. Pratt to Chris Dansberg, June 19, 1896; and letter from A. C. Pratt to John G. Taylor, June 25, 1896.

17. Mills, *Sagebrush Saga.* Also see Townley, "Reclamation in Nevada."

18. Statutes of the State of Nevada passed at the fourteenth session of the legislature, begun on Monday, January 7, and ended on Thursday, March 7 (Carson City: State Printer, 1889).

19. Townley in "Reclamation in Nevada" cites the *Belmont Courier*, August 10, 1889, p. 3, for an article on the necessity for smaller ranchers to sell to larger ranchers.

20. Townley, "Reclamation in Nevada."

21. "Biennial Report of the Surveyor-General and State Land Register, 1901–1902," in *Appendix to Journals of Senate and Assembly,* twenty-first session of the Nevada legislature (Carson City: State Printer, 1903). See also Statutes of the

State of Nevada passed at the nineteenth session of the legislature, commenced on Monday, January 16, and ended on Friday, March 10 (Carson City: State Printer, 1899).

22. Stewart, "The History of Range Use," 119–33. See also Hibbard, *History of Public Land Policies*; and Osgood, *The Day of the Cattleman.*

23. Frink et al., *When Grass was King*, 1–24.

24. Wooten, *Public Domain of Nevada.*

6. JOHN SPARKS: CAPITAL, CREDIT, AND COURAGE

1. Dictation by John Sparks on file at Bancroft Oral History Program, Bancroft Library, University of California, Berkeley.

2. R. G. Lillard, *Desert Challenge* (New York: Alfred A. Knopf, 1944). This description sounds like John Sparks, and one wonders if the author, who neither cites nor mentions an original source for the information, has confused the backgrounds of the partners.

3. Scarbrough, *Land of Good Water.*

4. Letter from Mrs. H. G. Scoggins to Mrs. J. B. Golden, June 26, 1979. Copy of letter furnished by Mrs. Golden.

5. *Elko Weekly Independent*, November 6, 1881.

6. *Elko Weekly Independent*, June 15, 1883.

7. Guinn, *History of the State of California.* Many modern accounts of the sale mention the formation of a million-dollar stock company to finance the purchase. There are no contemporary records of this company. Writers may have been confused by the million-dollar stock company formed by Sparks and Harrell in 1891.

8. *Elko Free Press*, September 1, 1883; *Weekly Drover's Journal* (Chicago), September 6, 1883. Also see letter to *Chicago Tribune*, July 7, 1883, quoting Beecher, in *Breeder's Journal* 4 (August 1883): 459.

9. Clay, *My Life on the Range.*

10. Truett, *On the Hoof and Horn in Nevada.*

11. C. A. Brennen, assisted by C. E. Fleming, G. H. Smith Jr., and M. R. Bruce, *Receipts and Costs on the Nevada Range Cattle Ranches for the Years 1928, 1929, and 1930*, Nev. Agric. Exp. Sta. Bull. 126 (Reno: University of Nevada, 1932).

12. Osgood, *The Day of the Cattleman.*

13. C. W. Hodgson, *Idaho Range Cattle Industry* (Moscow, Ida.: Animal Science Department, University of Idaho, 1948).

14. Truett, *On the Hoof and Horn in Nevada.*

15. R. R. Elliot, *History of Nevada* (Lincoln: University of Nebraska Press, 1973).

16. V. O. Goodwin, "Lewis Rice Bradley: Pioneer Nevada Cattleman and Nevada's First Cowboy Governor," *Nev. Hist. Soc. Q.* 14 (1977): 11–12. Goodwin refers to these cattle as "Texas Longhorns." There is little or no evidence that Texas Longhorns were regularly reaching Missouri in 1852. These were probably native American cattle or work oxen. Taking beef cattle to California in 1852 was akin to carrying coals to Newcastle.

17. Ibid. Goodwin says that Bradley moved to Nevada because of floods; however, Myron Angel, ed., in *History of Nevada* (Berkeley: Howell-North, reprint edition, 1881), says that Bradley moved because of drought. The same version appears in Buster King, "History of Lander County" (master's thesis, University of Nevada, Reno, 1954).

18. E. F. Treadwell, *The Cattle King* (New York: Macmillan, 1931).

19. *Humboldt Register,* January 15, 1870.

20. Townley, "Reclamation in Nevada."

21. C. B. Creel, *A History of Nevada Agriculture* (Reno: Max C. Fleischmann College of Agriculture, University of Nevada, 1964).

22. Gordon, "Report on Cattle, Sheep and Swine."

23. Clay, *My Life on the Range.*

24. *Elko Weekly Independent,* October 23, 1887.

25. *Daily Nevada State Journal,* July 15, 1885.

26. O. B. Peake, *The Colorado Range Cattle Industry* (Glendale, Calif.: Arthur H. Clark, 1937).

27. H. Goding and H. J. Raub, *The 28 Hour Law Regulating Interstate Transportation of Livestock: Its Purpose, Requirements, and Enforcement,* USDA Bull. 489 (Washington, D.C.: GPO, 1918).

28. J. R. Williams, *Cowboys out Our Way* (New York: Charles Scribner and Sons, 1951).

29. Gordon, "Report on Cattle, Sheep and Swine."

30. Ibid.

31. Truett, *On the Hoof and Horn in Nevada.*

32. Patterson et al., *Nevada's Northeast Frontier.*

33. Ibid.

34. Gordon, "Report on Cattle, Sheep and Swine."

35. Nimmo Report.

36. Ibid.

37. Frink, *Cow Country Cavalcade.*

38. Gordon, "Report on Cattle, Sheep and Swine."

39. Patterson et al., *Nevada's Northeast Frontier.*

40. *Carson Morning Appeal,* December 4, 1888.

41. Yost, *The Call of the Range.*

42. Stewart, "The History of Range Use."

43. W. M. Raine and W. C. Barnes, *Cattle* (New York: Grosset and Dunlap, 1930).

44. Personal communication to James A. Young from Newton J. Harrell, the son of Louis Harrell, Twin Falls, Ida., July 1979. After the accident, Louis Harrell rode by himself for two days to Montello, Nevada, where he boarded the train for Ogden, Utah. The hospital at Ogden succeeded in saving the sight of his other eye. While recovering in Ogden, Louis Harrell walked down to Browning Brothers Gunsmiths and purchased a new Colt single-action revolver. The revolver, original box, and receipt are the possessions of Newton Harrell.

45. Hawes, *The Valley of Tall Grass.*

46. Peake, *The Colorado Range Cattle Industry.*

47. R. F. Adams, *Come and Get It* (Norman: University of Oklahoma Press, 1952).

48. E. E. Dale, "Cowboy Cookery," *Amer. Hereford J.* 36, no. 17 (1946): 37–38.

49. Peake, *The Colorado Range Cattle Industry.*

50. As quoted in Frink et al., *When Grass Was King.*

51. John Sparks's dictation to Bancroft, on file at Bancroft Library, University of California, Berkeley.

7. WHITE WINTER: WHITE HILLS OF BONES

1. Dobie, *Longhorns.* Although primarily known for colorful descriptions of Longhorns, this volume contains valuable ecological material in the bibliographical notes.

2. The Nimmo Report is Ex. Doc. 267, 48th Cong., 2nd sess., House of Representatives (Washington, D.C.: GPO, 1885). It was prepared by Joseph Nimmo Jr., chief of the U.S. Bureau of Statistics.

3. Patents for barbed wire and machines for making the wire were granted to J. F. Clidden of De Kalb, Illinois, in 1874. See Osgood, *The Day of the Cattleman.*

4. The story of the accidental discovery that oxen could winter on the plains of Wyoming is in both the Nimmo Report and John Clay's *My Life on the Range.* Osgood considered the story pure fraud designed to enhance the stock prospectus of ranching companies. However, Agnes Wright Spring gives a detailed description of the event in Frick et al., *When Grass Was King,* and dates it to 1857. See the letter written by Alex Street, superintendent of Wells Fargo's freighting department at Cheyenne, Wyoming, and published by Dr. Hiram Latham in *Trans-Missouri Stock Raising: The Pasture Lands of North America: Winter Grazing* (Omaha: Nebraska Daily Herald Steam Printing House, 1871; reprint, Denver: Old West Publishing, 1962). Latham reported that he had wintered work oxen in Wyoming with no supplemental feed from 1856 to 1868. In about 1853, Seth E. Ward of Fort Laramie began to winter cattle in the valleys of the Chugwater and Laramie Rivers, according to Louis Petzer in *The Cattleman's Frontier* (Glendale, Calif.: Arthur H. Clark, 1936).

5. J. A. Young, R. E. Eckert Jr., and R. A. Evans, "Historical Perspective on the Sagebrush Grasslands," in *Proceedings of the Sagebrush Ecosystem: A Symposium* (Logan: Utah State University, 1978). Information on livestock herding supplied by J. H. Robertson, Professor Emeritus, Range Science, University of Nevada, Reno.

6. Latham published the first general appraisal of livestock raising on the open range. His pamphlet predated Joseph G. McCoy's *Historic Sketches of the Cattle Trade of the West and Southwest,* ed. Ralph Bicher (Glendale, Calif.: Arthur H. Clark, 1940), by three years.

7. J. A. Larson, "The Winter of 1886–87 in Wyoming," *Ann. Wyo.* 14 (1942): 5–18. See also Granville Stuart, *Forty Years on the Frontier,* ed. D. C. Phillips, 2 vols. (Glendale, Calif.: Arthur H. Clark, 1925).

8. The story of drifting cattle is from Osgood, *The Day of the Cattleman.* The cattle drifting in the streets is from Robert Fletcher, "The Hard Winter in Montana 1886–1887," *Agric. Hist.* 4, no. 4 (1930): 123–30. The newspaper accounts

are from R. H. Mattison, "The Hard Winter and the Range Cattle Business," *Montana: The Magazine of Western History* 1, no. 4 (1951): 5–22.

9. Bismarck weather data based on L. F. Crawford, *History of North Dakota* (Chicago: University of Chicago Press, 1931), 1:484.

10. Harold McCracken, *The Charles M. Russell Book* (Garden City, N.Y.: Doubleday, 1951).

11. Larson, "Winter of 1886–87 in Wyoming."

12. Frink et al., *When Grass was King*. This citation emphasizes the first section in the volume.

13. This comparison was probably first published by Osgood in *The Day of the Cattleman*. It has been noted by many authors; see, for example, Frink et al., *When Grass was King*. Larson, in "Winter of 1886–87 in Wyoming," gives Chicago market prices for September 6, 1887, as $2.75–3.45 per cwt, and $2.40–3.50 per cwt on November 8, 1887.

14. Gordon, "Report on Cattle, Sheep and Swine." The story of wintering cattle in Ruby Valley is from Petzer, *The Cattleman's Frontier*.

15. Ernest Antevs, *Rainfall and Tree Growth in the Great Basin* (Washington, D.C.: Carnegie Institute, 1938). Most of the tree-ring data are from the margins of the Great Basin.

16. Creel, *History of Nevada Agriculture*. See also Gordon, "Report on Cattle, Sheep and Swine." Gordon interviewed seventy cattlemen, who indicated losses of 20 percent.

17. Gordon, "Report on Cattle, Sheep and Swine." This report was prepared in the early 1880s, when the events of the winter of 1879–80 were relatively fresh in the minds of the stockmen interviewed.

18. Hodgson, *Idaho Range Cattle Industry*. A copy of this rare 1948 report was supplied by the Idaho Historical Society.

19. Patterson et al., *Nevada's Northeast Frontier*, contains a great deal of source material collected by Mrs. Patterson during a lifetime in Elko County. It is further enriched by the contributions of the other authors.

20. A. E. Granger, M. M. Bill, G. C. Simmons, and F. Lee, *Geology and Mineral Resources of Elko County, Nevada*, Nev. Bur. Mines Bull. 54 (Reno: University of Nevada, 1957). Long-term precipitation averages are compiled in this bulletin. Newspaper accounts are from Patterson et al., *Nevada's Northeast Frontier*.

21. V. S. Truett, "White Winter of 89 and 90," *Nevada Magazine* 3 (January 1948): 5–9, 28. This article provides the basis for many of the comments about the midwinter period in Nevada. Quote from *Elko Weekly Independent*, December 12, 1889.

22. *Elko Independent*, December 22 and 29, 1889.

23. *Elko Independent*, January 12, 1890.

24. *Elko Independent*, February 2, 1890.

25. *Elko Independent*, February 16, 1890.

26. *Reno Evening Gazette*, February 10 and 20, 1890.

27. Byrd Trego's article about the hard winter of 1889–90 published in the *Idaho Republic* of May 17, 1928, is one of the best first-person accounts of that winter.

28. Reports of the hard winter of 1889–90 are from J. D. Oliphant, *On the Cattle Ranges of the Oregon Country* (Seattle: University of Washington Press, 1968). On page 284 of this volume, Oliphant provides newspaper citations for documentation of winter losses from each area mentioned.

29. Treadwell, *The Cattle King*.

30. H. O. Brayer, "The L-7 Ranch: An Incident of the Economic Development of the Western Cattle Industry," *Ann. Wyo.* 15 (January 1943): 5–37.

31. Hawes, *The Valley of Tall Grass*. The author confuses the winter of 1889–90 with the winter of 1888, but the volume contains much valuable source material.

32. *Elko Independent*, March 16, 1890. Many knowledgeable residents of Elko County question the often-quoted story of stepping from carcass to carcass along the Marys River. The Marys River is not one hundred miles long, and many people believe the story refers to the Salmon River and Sparks-Tinnin rangeland.

33. *Elko Independent*, March 2, 1890.

34. C. S. Walgamott, *Six Decades Back* (Caldwell, Ida.: Caxton, 1936). This book is believed to be the published origin of the statistic, which Walgamott probably obtained from folklore, newspaper accounts, and word-of-mouth accounts of the era.

35. *Harper's Weekly* 47 (June 20, 1902): 1017–34. Hereafter cited as *Harper's Weekly*, "Nevada."

36. This version of the meeting was first published in Walgamott, *Six Decades Back*. Tinnin did go to Nebraska, where he continued to raise Longhorns; see Yost, *The Call of the Range*.

37. The $45,000 price is for one-half or one-third interest in a $1 million stock company. The incorporation papers were filed with the Office of the Secretary of State of California on February 12, 1891.

38. R. H. Lane, "The Cultural Ecology of Sheep Nomadism: Northeastern Nevada, 1870–1972" (Ph.D. diss., Yale University, New Haven, 1974).

39. J. Budy and J. A. Young, "Historical Uses of Pinyon Juniper Woodlands," *Forest Hist.* 23, no. 3 (1979): 113–21.

8. WATER: THE FINITE RESOURCE

1. The development of water rights on Goose Creek is reviewed in the records of the District Court of the United States for Idaho, Southern Division, for the case of *Twin Falls-Oakley Land and Water Company and Oakley Land Company* v. *Vineyard Land and Stock Company*, August 14, 1915, transcript on file with Nevada State Engineer's Office, Carson City. Jedediah Smith visited Salmon Falls Creek in 1826 but failed to find beaver. Milton Sublette and a party of trappers from the Rocky Mountain Fur Company trapped beaver along Goose Creek in 1837. See D. L. Morgan, *Jedediah Smith and the Opening of the West* (Lincoln: University of Nebraska Press, 1953).

2. D. W. Thorne, "The Desert Shall Blossom As the Rose," Tenth Annual Faculty Research Lecture (Logan: Faculty Association, Utah State Agricultural College, 1951).

3. One of the early books on irrigation that mentions irrigation by Mormons is F. H. Newell and D. W. Murphy, *Principles of Irrigation Engineering* (New York: McGraw-Hill, 1913). Comments on Indian irrigation are from J. H. Steward, *Basin-Plateau Aboriginal Socio-political Groups* (Washington, D.C.: GPO, 1938).

4. A. Steinel, *History of Agriculture in Colorado* (Fort Collins: Colorado State Agricultural College, 1926).

5. A. H. Sanford, *The Story of Agriculture in the United States* (Boston: D. C. Heath, 1961).

6. T. A. Garrity and E. Nitzschke Jr., *Water Law Atlas*, N. Mex. Inst. Mining Technol. Circ. 95 (Socorro: New Mexico Chapter of the Federal Bar Association in cooperation with the State Bureau of Mines and Mineral Resources, 1957).

This atlas is an excellent and simple presentation of the basic principles of water law.

7. G. Dangberg, *Conflict on the Carson* (Minden, Nev.: Carson Valley Historical Society, 1976). See also Wooton, *The Public Domain of Nevada.*

8. Townley, "Reclamation in Nevada." See also the reproduction of the 1881 edition of Thompson and West's *History of Nevada,* ed. M. Angel, with illustrations and biographical sketches of its prominent men and pioneers (Berkeley: Howell-North, 1958).

9. *Biennial Report of the State Engineers, State of Nevada, June 30, 1939 to July 1, 1940* (Carson City: State Printing Office, 1940).

10. Records of the District Court of the Fourth Judicial District of the State of Nevada in and for the County of Elko, Judgment and decree of the matter of the determination of the relative rights in and to the water of the Salmon River, March 1922. See also Fulton, "Camp Life on Great Cattle Ranches." Fulton is often misquoted as saying that Sparks harvested 100,000 tons of hay.

11. Adelaide Hawes's 1950 book *The Valley of Tall Grass* lists the priority of all the water rights on the Bruneau River. The history of irrigation districts is from Sanford, *The Story of Agriculture.*

12. The Alvaro Evans family's papers are on file at the Nevada Historical Society in Reno. The item is dated June 1879.

13. C. A. Brennen, assisted by C. E. Fleming, G. H. Smith Jr., and M. R. Bruce, *Cost of Producing Hay on Nevada Ranches,* Nev. Agric. Exp. Sta. Bull. 129 (Reno: University of Nevada, 1932).

14. Nimmo Report.

15. Comments on desert soils and water quality are largely based on Thorne, "The Desert Shall Blossom As the Rose."

16. *Reno Evening Gazette,* October 16, 1880.

17. R. E. Rush, *Water Resources,* Reconnaissance Ser. Rep. 47, Department of Conservation and Natural Resources (Carson City: State of Nevada, 1968). Water resource appraisal of Thousand Springs Valley, Elko County, Nevada.

18. B. H. Hibbard, *History of Public Land Policies*; and *Statutes and Regulations Governing Entries and Proof under the Desert Land Laws,* Gen. Land Off. Circ. 474, USDI (Washington, D.C.: GPO, 1916).

19. Hibbard, *History of Public Land Policies.*

9. MAKING HAY IN THE GREAT BASIN

1. The quote attributed to a disgruntled cowboy is from Osgood, *The Day of the Cattleman.*

2. Statistics on "male" Nevada are from the burning social commentary of Anne Martin, "Nevada, Beautiful Desert of Buried Hopes," *Nation* 115 (July 26, 1922): 89–92.

3. Nevada politics is outside the limits of this volume. For more on this subject, see G. L. Ostrander, *Nevada: The Great Rotten Borough, 1859–1864* (New York: Alfred A. Knopf, 1966).

4. H. L. Davis, *Winds of Morning* (New York: William Morrow, 1952).

5. The dirty plate rules are in Treadwell, *The Cattle King.*

6. J. K. Woodfield, "The Initiation of the Range Cattle Industry of Utah" (master's thesis, University of Utah, Salt Lake City, 1957).

7. *Elko Weekly Independent,* April 17, 1887.

8. Townley, "Reclamation in Nevada."

9. Alvaro Evans Papers, Nevada Historical Society, Reno.

10. Townley, "Reclamation in Nevada."

11. *Reno Crescent,* February 19, 1874.

12. *Reno Crescent,* September 16, 1871; *Nevada State Journal,* May 24, 1873.

13. Townley, "Reclamation in Nevada."

14. David Griffiths, *Forage Conditions on the Northern Border of the Great Basin,* USDA Bur. Plant Industry Bull. 15 (Washington, D.C.: GPO, 1902).

15. Professor Frank Lamson-Scribner's comments are cited in *Range Plant Handbook.*

16. Creeping wildrye often occurs on saline soils. Griffiths had trouble distinguishing saline/alkaline from nonsalty soils.

17. J. McCormick, J. A. Young, and W. Burkhardt, "Making Hay," *Rangelands* 1, no. 5 (1979): 203–6.

18. G. Stewart, *Alfalfa Growing in the United States and Canada* (New York: Macmillan, 1926).

19. Henry Miller is credited with being the first to extensively grow alfalfa, cotton, and rice in California. See Treadwell, *The Cattle King.*

20. M. Angel, ed., *History of Nevada; Humboldt Register,* April 2, 1864, p. 2; *Daily Nevada State Journal,* September 25, 1874, p. 2, and May 7, 1876, p. 2.

21. "History of Alfalfa Cultivations in Nevada," in C. A. Norcross, *Alfalfa and Sweet Clover*, Agric. Ext. Bull. 5 (Reno: University of Nevada, 1916). The Fred Dangberg information is from Lillard, *Desert Challenge*.

22. For references to ranch improvements by Sparks, see Patterson et al., *Nevada's Northeast Frontier*. J. A. Graves saw alfalfa in 1896; see Graves, *California Memories* (Los Angeles: Times-Mirror Press, 1930).

23. Stewart, *Alfalfa Growing*.

24. Samuel Fortier, *Irrigation of Alfalfa*, USDA Farmer's Bull. 373 (Washington, D.C.: GPO, 1909).

25. C. M. Saine, "Alfalfa in Nevada's Lovelock Valley," *Sunset* 7 (June–July 1901): 31-32.

26. For an economist's point of view, see Wooten, *The Public Domain of Nevada*; for social commentary, see Martin, "Beautiful Desert."

27. MacDonald, *Food from the Far West*.

28. Information on old mower is from Mills, *Sagebrush Saga*. Disassembly of mower from personal communication with Al and Norman Glaser of Halleck, Nevada. Cost of haying implements from Brennen et al., *Cost of Producing Hay on Nevada Ranches*. Information on reapers from A. H. Sanford, *The Story of Agriculture in the United States* (Boston: D. C. Heath, 1916).

29. Bowman, *Only the Mountains Remain*.

30. Personal communication to J. A. Young from Josephine S. Barterelli, Battle Mountain, Nev., whose father followed this practice of teaming wild and gentle workhorses on mowers. Comments on handling green horses are personal communication to J. A. Young from Donna Anderson.

31. H. L. Davis, *Honey in the Horn* (New York: William Morrow, 1938). This novel, which won a Pulitzer Prize for fiction, contains excellent descriptions of hay camps.

32. Owner's manual for Meadow King mower, manufactured in 1870 by Gregg, Plyer and Company, Trumansburgh, Tompkins County, New York.

33. L. W. Fluharty and J. C. Hays, *Wild-Hay Management Practices in Modoc County*, Univ. Calif. Agric. Exp. Sta. Bull. 679 (Davis: University of California, 1943).

34. In some farming areas, the term "haycock" was universally used; in others, "shocks" was the only term used.

35. Personal communication to J. A. Young from Donna Anderson.

36. H. E. Selby, *Cost and Efficiency in Producing Alfalfa Hay in Oregon* (Corvallis: Oregon State Agricultural College Experiment Station, 1928).

37. Stewart, *Alfalfa Growing.*

38. "Sanitary Bill" was the exact opposite of what his name implied. See Bowman, *Only the Mountains Remain.*

39. David Griffiths watched and described a slide work at the Island Ranch near Burns, Oregon, in 1901. See Griffiths, *Forage Conditions.*

40. Description of beaverslide prepared by Dr. Wayne Burkhardt, Assistant Professor of Range Science, University of Nevada, Reno; description of Elko beaverslides provided by Donna Anderson.

41. Griffiths, *Forage Conditions.*

42. Personal communication to J. A. Young from Virgil Knutson, Adin, Calif.

10. FROM DUGOUTS TO CATTLE EMPIRES

1. The description of a typical nineteenth-century ranch headquarters is from Mrs. Agnes Robison, talking about the McGill Ranch in White Pine County, Nevada. See also Orville Holderman, "Jewett W. Adams and W. M. McGill: Their Lives and Ranching Empire" (master's thesis, University of Nevada, Reno, 1963).

2. F. C. Scharader, *A Reconnaissance of the Jarbidge, Contact, and Elko Mountain Mining Districts, Elko County, Nevada,* USGS Bull. 497 (Washington, D.C.: GPO, 1912).

3. M. F. Knudtsen, *Here Is Our Valley,* Helen Marye Thomas Mem. Ser. No. 1 (Reno: Max C. Fleischmann College of Agriculture, University of Nevada, 1975).

4. Personal communication to J. A. Young from old-time horseshoer Bill Smith, Etna, Calif.

5. Graves, *California Memories.* Years later, this custom-made buggy was found behind a store in Montello and restored by the Bowman family.

6. Hawes, *The Valley of Tall Grass.*

7. Knudtsen, *Here Is Our Valley.* Sparks at one time had a contract with the U.S. government to deliver mail from Montello up Thousand Springs Valley and down Salmon Falls Creek into Idaho. Essentially, the government paid the expenses of shipping supplies to the Sparks-Harrell ranches.

8. Brennen et al., *Cost of Producing Hay on Nevada Ranches.*

9. Treadwell, *The Cattle King.*

10. Raine and Barnes, *Cattle.*

11. Knudtsen, *Here Is Our Valley.*

12. Brennen et al., *Cost of Producing Hay on Nevada Ranches.*

13. Graves, *California Memories.*

14. Mills, *Sagebrush Saga.*

11. HEREFORDS IN THE SAGEBRUSH

1. MacDonald, *Food from the Far West.*

2. D. F. Malin, *The Evolution of Breeds* (Des Moines, Iowa: Wallace Publishing, 1923).

3. Ibid.

4. Ralph Bogart, *Improvement of Livestock* (New York: Macmillan, 1959).

5. Malin, *Evolution of Breeds.*

6. Ibid.

7. *Elko Independent,* November 14, 1874.

8. James Sinclair, *History of Shorthorn Cattle* (London: Vinton, 1907).

9. MacDonald, *Food from the Far West.*

10. Patterson et al., *Nevada's Northeast Frontier.*

11. G. M. Rommel, *American Breeds of Cattle with Remarks on Pedigrees,* U.S. Bur. Animal Industry Bull. 34 (Washington, D.C.: GPO, 1902).

12. Malin, *Evolution of Breeds.*

13. Ibid.

14. J. M. Hazelton, *A History of Linebreed Anxiety 4th Herefords of Straight Gudgell and Simpson Breeding* (Kansas City, Mo.: Association of Breeders of Anxiety 4th Herefords, 1939).

15. H. Bancroft, *History of Nevada, Colorado, and Wyoming* (San Francisco: History, 1890).

16. A. H. Sanders, "The Story of Herefords," *Breeder's Gazette,* Chicago, 1914. See also *Weekly Drover's Journal* (Chicago), August 28, 1879.

17. Patterson et al., *Nevada's Northeast Frontier.*

18. Creel, *History of Nevada Agriculture.*

19. Holderman, "Jewett W. Adams and W. M. McGill."

20. *Elko Independent,* August 19, 1885.

21. Statement prepared by Phillip Earl in support of making the Sparks mansion a historical landmark. Material on file at Nevada Historical Society, Reno.

22. *Harper's Weekly*, "Nevada."

23. Ibid. Several works such as Patterson's *Nevada's Northeast Frontier* give dates in the 1890s for the start of Sparks's registered Hereford venture.

24. *Pacific Rural Press*, January 5, 1889, p. 5.

25. V. S. Truett, "From Longhorns to Herefords: A Pioneer in Nevada's Cattle Industry," and "John Sparks, Cattleman Who Became Governor," *Nevada Highways and Parks*, Special Centennial Issue (1964): 30–31, 50–63.

26. Hazelton, *History of Linebreed Anxiety 4th Herefords*. See also *Reno Evening Gazette*, November 25, 1895.

27. Sanders, "Story of Herefords."

28. *Reno Evening Gazette*, March 4, 1898.

29. *Reno Evening Gazette*, April 31, 1896.

30. Sam Davis, "The Governor of Nevada: Some Characteristics of the Chief Executive of the Silver State," *Sunset* (May 1903): 19–20.

31. *Reno Evening Gazette*, November 25, 1895.

32. *Reno Evening Gazette*, May 1, 1897.

33. S. H. Short, "A History of the Nevada Livestock Industry Prior to 1900" (master's thesis, University of Nevada, Reno, 1965).

34. D. R. Ornduff, *The Hereford in America: A Compilation of Historic Facts about the Breed's Background and Bloodlines* (Kansas City, Mo.: Hereford History Press, n.d.).

35. *Harper's Weekly*, "Nevada."

36. For examples, see Truett, "From Longhorns to Herefords"; and Sparks's obituary, on file at Nevada Historical Society, Reno. See also Davis, "Governor of Nevada."

37. Malin, *Evolution of Breeds*.

38. Sanders, "Story of Herefords."

39. Information on the purchase of Herefords by John Sparks supplied by D. Prothero, Hereford Herd Book Society, Hereford House, Hereford, England. Sources include the Expert Ledger for 1897 and Auctioneer's Remarks for the sale of the Leen, Courthouse, and Moukron herds.

40. Davis, "Governor of Nevada."

41. *Harper's Weekly*, "Nevada."

42. Truett, "Longhorns to Herefords."

43. Sparks's obituary, Nevada Historical Society, Reno.

44. Truett, "Longhorns to Herefords."

45. *Nevada State Journal*, July 28, 1900.

46. Short, "History of the Nevada Livestock Industry"; and Davis, "Governor of Nevada."

47. Sanders, "Story of Herefords"; and Truett, "Longhorns to Herefords." Armour Rose arrived in Reno on December 18, 1899. *Reno Evening Gazette*, December 18, 1899.

48. Truett, "Longhorns to Herefords."

49. R. I. Fulton, "Camp Life on Great Cattle Ranches."

50. Sanders, "Story of Herefords."

51. *Reno Evening Gazette*, October 28, 1899.

52. *Harper's Weekly*, "Nevada."

53. Sanders, "History of Herefords."

54. E. G. Ritzman, "The Development of Livestock Shows and Their Influences on Cattle Breeding and Feeding," in *U.S. Bureau of Animal Industry Annual Report* 25:245–56 (Washington, D.C.: GPO, 1908).

12. HORSES, TAME AND WILD IN THE SAGEBRUSH

1. T. McKnight, *Feral Livestock in Anglo-America*, Univ. Calif. Publ. Geogr. 16 (Berkeley: University of California Press, 1964). Camels were allowed to become feral in the Carson Desert during the late nineteenth century, but breeding populations never became established. The animals were probably too old to breed when allowed to roam free.

2. D. I. Axelred, "Evolution of Desert Vegetation in Western North America," in *Contributions to Paleontology Studies in Late Tertiary Paleobotany*, Carnegie Inst. Wash. Publ. 590, 217–99, (Washington D.C.: Carnegie Institute, 1950).

3. E. H. Cronin, P. Ogden, J. A. Young, and W. Laycock, "The Ecological Niches of Poisonous Plants in Range Communities," *J. Range Manage.* 31 (1978): 328–34.

4. P. S. Martin and H. E. Wright Jr., eds., *Pleistocene Extinctions*, vol. 6 of *Proceedings of the VIII Congress of the International Association for Quaternary Research* (New Haven: Yale University Press, 1967).

5. B. M. Underhill, "The Evolution of the Horse," *Sci. Amer.* 64 (1907): 413.

6. W. E. Jones and R. Bogart, *Genetics of the Horse* (East Lansing, Mich.: Caballus, 1971).

7. Ibid.

8. W. D. Wyman, *The Wild Horses of the West* (Lincoln: University of Nebraska Press, 1945).

9. Jones and Bogart, *Genetics of the Horse.*

10. Clark Wissler, "The Influence of the Horse in the Development of Plains Culture," *Amer. Anthropol.* 16 (1914): 24–32.

11. Francis Haines, "The Northward Spread of Horses among the Plains Indians," *Amer. Anthropol.* (July–September 1938): 141–53.

12. H. H. Bancroft, *California Pastoral* (San Francisco: History, 1888).

13. H. Bidwell, *Echoes of the Past* (New York: Citadel Press, 1962).

14. Dan DeQuille, *Washoe Rambles* (Los Angeles: Westernlore Press, 1963).

15. J. H. Steward, *Basin-Plateau Aboriginal Sociopolitical Groups* (Washington D.C.: GPO, 1938).

16. Captain J. H. Simpson, *Report of Explorations across the Great Basin in Utah in 1859* (Washington, D.C. GPO, 1876).

17. Morgan, *The Humboldt Highroad of the West.*

18. Truett, *On the Hoof and Horn in Nevada.*

19. Patterson et al., *Nevada's Northeast Frontier.*

20. Gordon, "Report on Cattle, Sheep and Swine."

21. *Elko Independent,* April 10, 1887.

22. Mills, *Sagebrush Saga.*

23. Brennen et al., *Cost of Producing Hay on Nevada Ranches.*

24. R. Steele, *Mustangs on the Mesas* (Los Angeles: Mabel Walden Steele, 1941).

25. Charles "Pete" Barnum as cited in ibid.

26. Steele, *Mustangs on the Mesas.*

27. Personal communication from Melvin James Sharp, formerly of Nye County, Nev.

28. Personal communication from Thomas R. Bunch, formerly of Burnt River, Ore. To this day Old Bunch breaks out in a rash if served applesauce.

29. R. Steele, "Trapping Wildhorses in Nevada," *McClures* 24, no. 2 (December 1909): 198–209.

30. S. W. Pellegrini, "Home Range, Territoriality and Movement Patterns of

Wild Horses in the Wassuk Range of Western Nevada" (master's thesis, University of Nevada, Reno, 1971).

31. Bowman, *Only the Mountains Remain.*

32. F. Robbins, "Capturing Wild Horses," *Western Horseman* (November–December 1947): 13–15, 48.

33. A. Amaral, *Mustang* (Reno: University of Nevada Press, 1977).

34. Steele, *Mustangs on the Mesas.*

35. Davis, *Winds of Morning.*

36. Steele, *Mustangs on the Mesas;* Amaral, *Mustang.*

37. Pellegrini, "Home Range, Territoriality and Movement Patterns of Wild Horses"; personal communication from cheerful Charlie Fisher, Carson City, Nev.

38. Pellegrini, "Home Range, Territoriality and Movement Patterns of Wild Horses."

39. Bogart, *Improvement of Livestock.*

40. Jones and Bogart, *Genetics of the Horse.*

41. Bogart, *Improvement of Livestock.*

42. Ibid.

43. R. H. MacArthur and J. H. Connell, *The Biology of Populations* (New York: John Wiley and Sons, 1956), and R. H. MacArthur and E. O. Williams, *The Theory of Island Biogeography* (Princeton: Princeton University Press, 1967), are good sources on population genetics.

44. Alvaro Evans Papers, Nevada Historical Society, Reno.

45. *Reno Evening Gazette,* March 19, 1985.

46. *Reno Evening Gazette,* April 29, 1885.

47. Amaral, *Mustang.*

48. Patterson et al., *Nevada's Northeast Frontier.*

49. Scarbrough, *Land of Good Water;* also B. C. Hanes, *Bill Pickett, Bulldogger: The Biography of a Black Cowboy* (Norman: University of Oklahoma Press, 1970).

50. Personal communication from N. T. Harrell, Twin Falls, Ida., July 1979.

51. Hawes, *The Valley of Tall Grass.*

52. Personal communication from N. T. Harrell, Twin Falls, Ida., July 1979.

53. Ibid.

54. Death certificate from Janet M. Wick, State Registrar, State of Idaho,

Department of Health and Welfare, Boise. Henry died April 3, 1937, in Twin Falls. His occupation was listed as ranch foreman with residence at San Jacinto, Nevada.

55. Personal communication from N. T. Harrell, Twin Falls, Ida., July 1979.

56. D. O. Hyde, *The Last Free Man* (New York: Dial Press, 1973).

57. *Elko Weekly Independent,* September 11, 1887: "John Sparks of Sparks and Tinnin left Wells for his holding with a pair of thoroughbred foxhounds. No one on the Pacific Coast can equal Sparks's dogs."

58. Personal communication from Ms. Lillian Pepper of Dayton, Nev.

59. *Reno Evening Gazette,* January 25, 1890.

60. *Reno Evening Gazette,* August 3, 1895.

61. Graves, *California Memories.*

62. *Wadsworth Dispatch,* November 15, 1893.

63. *Silver State,* February 23, 1899.

13. THE PASSING OF THE OLD GUARD

1. Letter to Louis Harrell dated February 12, 1895, and signed by John Sparks, president, and A. J. Harrell, secretary, Sparks-Harrell Company, Visalia, Calif. Copy of letter furnished by Newton T. Harrell, Twin Falls, Ida.

2. Patterson et al., *Nevada's Northeast Frontier.*

3. Letter to Louis Harrell dated December 28, 1896, and signed by A. J. Harrell, Visalia, Calif. Copy of letter furnished by Newton T. Harrell, Twin Falls, Ida. Harris and Duncan were Henry Harris and Tap Duncan, the Sparks-Harrell wagon or roundup foremen. The desert is probably the Owyhee Desert of south-central Idaho.

4. Byrd Fanita Wall Sawyer, *Nevada Nomads* (San Jose, Calif.: Harlan-Young Press, 1971).

5. Most of the information on Diamondfield Jack is from *Diamondfield Jack,* by D. H. Grover (Reno: University of Nevada Press, 1968). Materials collected from sources Grover did not use are cited as separate sources.

6. Truett, *On the Hoof and Horn in Nevada.* Information on Sparks's military career in Wyoming from H. H. Bancroft Collections, Bancroft Library, University of California, Berkeley.

7. Patterson et al., *Nevada's Northeast Frontier.*

8. Louis Harrell told his sons that he saw Diamondfield Jack arrive at the Boar's Nest Ranch and his horse did not appear to have been ridden hard. This was brought out in court by other witnesses.

9. Wooten, *The Public Domain of Nevada.* Wooten recognized the relationship among taxes, production, and landownership and the disproportionate tax load that agricultural landowners carried during the late nineteenth and early twentieth centuries.

10. To keep competition away from its cattle ranges, the Utah Construction Company fully occupied the available winter range with more than forty thousand ewes. See Bowman, *Only the Mountains Remain.*

11. Governor Sadler's papers are on file at the Nevada Historical Society, Reno.

12. Griffiths, *Forage Conditions.*

13. Fulton, "Camp Life on Great Cattle Ranches."

14. The *Reno Evening Gazette,* August 21, 1899, indicated that Sparks boarded the train in Reno for his annual hunting trip.

15. Fulton, "Camp Life on Great Cattle Ranches." B. Abbott Sparks Jr. located Alcade in California in 1984.

16. Graves, *California Memories.*

17. *Reno Evening Gazette,* June 3, 1895, July 3, 1895.

18. Governor Sadler's papers are on file at the Nevada Historical Society, Reno.

19. Ibid.

20. C. F. Martin, *Proceedings of the National Stock Growers Association Convention* (Denver: Smith-Brock Printing, 1898).

21. Creel, *History of Nevada Agriculture.*

22. W. H. Poulton was the granduncle of C. E. Poulton, a noted range scientist, educator, plant ecologist, and specialist in remote sensing.

23. Martin, *Proceedings of the National Stock Growers Convention.*

24. Hawes, *The Valley of Tall Grass.* Remarks about the end of the Sparks-Harrell operation on the Owyhee Desert are from a 1937 letter that Mrs. Hawes's father sent to the Regional Forester, USDA Forest Service, Ogden, Utah. Apparently Louis Harrell never organized the last big roundup as instructed by A. J. Harrell.

25. See Patterson et al., *Nevada's Northeast Frontier.*

26. Guinn, *History of the State of California.*

27. Graves, *California Memories.*

28. *Reno Evening Gazette,* April 4, 1901. Sparks already owned ten thousand acres in Texas; his will does not mention the additional twelve thousand acres.

29. In 1903, Elko County tax assessment rolls listed Sparks-Harrell as the third-largest taxpayer in Elko County with an assessed value of $242,425; *Wells State Herald,* September 18, 1903.

30. *Harper's Weekly,* "Nevada."

31. Little is known about John Sparks's financial status in Nevada when he died except the general knowledge that he was bankrupt. His estate in Texas listed assets of $47,015 and debts of $57,784 in Probate No. 1554, Williamson County, Texas, September 28, 1908.

32. Guinn, *History of the State of California.*

33. For information on late-nineteenth-century scientists who were active in describing the forage resources of the sagebrush/grasslands, see F. G. Renner, *A Selected Bibliography on Management of Western Ranges, Livestock, and Wildlife,* USDA Misc. Publ. 281 (Washington, D.C.: GPO, 1938).

34. *Wells State Herald,* August 15, 1905.

35. Nelson, *The Red Desert of Wyoming.*

36. John Sparks purchased the Wedekind mine for $175,000 in cash (a newspaper editor saw the money). Residents of Reno thought Wedekind was crazy because he kept prospecting in the ridge north of the city; see *Reno Evening Gazette,* May 31, 1902. It is remarkable that John Sparks, who had invested in mining ventures in Wyoming in the 1870s and in Nevada for twenty-five years, was victimized in the Wedekind mine fraud. Apparently, the surface deposits were rich oxides of silver and the mill was built to treat such ores; but at depth, the ores became sulfates that the mill could not treat. The mine also encountered a great deal of hot water. The area is a valid mineral prospect and Sparks probably suffered from bad luck and poor business judgment. See H. F. Bonham, *Geology and Mineral Deposits of Washoe and Storey Counties, Nevada,* Mackay School of Mines Bull. 70 (Reno: University of Nevada, 1969). After his death, Sparks's ranch in Texas became the site of the Texaco oil fields.

37. Clay, *My Life on the Range.*

38. *Water and Related Land Resources.*

EPILOGUE

1. During the 2000 fire season, pilots of aircraft used in fire suppression were lost.

2. J. A. Young, A. Evans, R. E. Eckert Jr., and B. L. Kay, "Cheatgrass," *Rangelands* 2 (1987): 266–71.

3. J. A. Young and D. McKenzie, "Rangeland Drill," *Rangelands* 4 (1982): 108–13.

EQUIVALENCY NAME TABLE

PLANTS

Abies lasiocarpa—subalpine fir

Agoseris glauca—pale agoseris

Agropyron smithii—western wheatgrass

Agropyron spicatum—bluebunch wheatgrass

Agrostis alba—redtop

Artemisia arbuscula—low sagebrush

Artemisia cana—silver sagebrush

Artemisia nova—black sagebrush

Artemisia spinescens—budsage

Artemisia tridentata ssp. *tridentata*—basin big sagebrush

Artemisia tridentata ssp. *vaseyana*—mountain big sagebrush

Artemisia tridentata ssp. *wyomingensis*—Wyoming big sagebrush

Atriplex—saltbush

Atriplex canescens—fourwing saltbush

Atriplex confertifolia—shadescale

Balsamorhiza sagittata—arrowleaf balsam root

Bromus tectorum—cheatgrass, downy brome

Carex sp.—tuff sod of sedges

Castilleja sp.—Indian paintbrush

Ceratoides lanata—winterfat

Cercocarpus ledifolius—curlleaf mountain mahogany

Chrysothamnus viscidiflorus—low rabbitbrush

Collinsia sp.—Chinese house

Cowania mexicana ssp. *stansburiana*—cliffrose

Dactylis glomerata—orchard grass

Delphinium sp.—low larkspur

Delphinium barbeyi—tall larkspur

Delphinium glaucum—tall larkspur

Elocharis aricularis—spike rush

Elymus cinereus—Great Basin wildrye

Elymus triticoides—diminutive creeping wildrye

Ephedra viridis—green ephedra

Festuca idahoensis—Idaho fescue

Gossypium sp.—cotton

Hordeum brachyantherum—meadow barley

Hordeum vulgare—barley

Iris missouriensis—Rocky Mountain iris

Juncus balticus—wiregrass

Juniperus osteosperma—Utah juniper

Medicago sativa—alfalfa

Mimulus nanus—skunk monkey flower

Muhlenbergia brachyantherum—mat muhly

Orzyopsis hymenoides—Indian ricegrass

Phleum pratense—timothy

Pinus albicaulis—white bark pine

Pinus contorta—lodgepole pine

Pinus flexilis—limber pine

Pinus longalva—ancient bristlecone

Pinus monophylla—single-leaf pinyon

Poa nevadensis—Nevada bluegrass

Poa sandbergii—Sandberg bluegrass

Populus fremontii—Fremont cottonwood

Populus nigar var. *italica*—Lombardy poplar

Populus tremuloides—quaking aspen

Populus trichocarpa—black cottonwood

Prunus andersonii—desert peach

Purshia tridentata—bitterbrush

Ribes velutinum—ribes

Salix sp.—willow

Sarcobatus vermiculatus—greasewood

Scirpus robustus—alkali bullrush

Sitanion hystrix—squirreltail
Stipa comata—needle and thread grass
Stipa thurberiana—Thurber needlegrass
Symphoricarpos sp.—snowberry
Tetradymia canescens—horsebrush
Typha latifolia—cattail tule
Vulpia octoflora—six-weeks fescue
Wyethia mollis—mule ears
Zigadenus venenosus—death camas

ANIMALS

Antilocapra americana—pronghorn
Bison bison—American bison
Bos—cow
Camelops—camel
Castor canadensis—beaver
Centrocerus urophasianus—sage grouse
Cervus canadensis—elk, wapiti
Equus caballus—horse
Gymnorhinus cyanocephala—pinyon jays
Lepus californicus—black-tailed jackrabbit
Lepus townsendii—white-tailed jackrabbit
Mastodon sp.—mastodon
Odocoileus hemionus—mule deer
Ovis aries—sheep
Ovis canadensis—desert bighorn
Pogonomyrmex sp.—harvester ants

BACTERIA

Bacillus tularense—tularemia